# Lecture Notes in Electrical Engineering

## Volume 819

The book series *Lecture Notes in Electrical Engineering* (LNEE) publishes the latest developments in Electrical Engineering - quickly, informally and in high quality. While original research reported in proceedings and monographs has traditionally formed the core of LNEE, we also encourage authors to submit books devoted to supporting student education and professional training in the various fields and applications areas of electrical engineering. The series cover classical and emerging topics concerning:

- Communication Engineering, Information Theory and Networks
- Electronics Engineering and Microelectronics
- Signal, Image and Speech Processing
- Wireless and Mobile Communication
- Circuits and Systems
- Energy Systems, Power Electronics and Electrical Machines
- Electro-optical Engineering
- Instrumentation Engineering
- Avionics Engineering
- Control Systems
- Internet-of-Things and Cybersecurity
- Biomedical Devices, MEMS and NEMS

For general information about this book series, comments or suggestions, please contact leontina.dicecco@springer.com.

To submit a proposal or request further information, please contact the Publishing Editor in your country:

**China**

Jasmine Dou, Editor (jasmine.dou@springer.com)

**India, Japan, Rest of Asia**

Swati Meherishi, Editorial Director (Swati.Meherishi@springer.com)

**Southeast Asia, Australia, New Zealand**

Ramesh Nath Premnath, Editor (ramesh.premnath@springernature.com)

**USA, Canada:**

Michael Luby, Senior Editor (michael.luby@springer.com)

**All other Countries:**

Leontina Di Cecco, Senior Editor (leontina.dicecco@springer.com)

**** This series is indexed by EI Compendex and Scopus databases. ****

More information about this series at https://link.springer.com/bookseries/7818

Shiban Kishen Koul · Zamir Wani

# Novel Millimetre Wave Antennas for MIMO and 5G Applications

Springer

Shiban Kishen Koul
Centre for Applied Research in Electronics
Indian Institute of Technology Delhi
New Delhi, India

Zamir Wani
Institute Innovation and Entrepreneurship
Development Centre (I2EDC)
Indian Institute of Technology Jammu
Jammu, India

ISSN 1876-1100 ISSN 1876-1119 (electronic)
Lecture Notes in Electrical Engineering
ISBN 978-981-16-7277-4 ISBN 978-981-16-7278-1 (eBook)
https://doi.org/10.1007/978-981-16-7278-1



This Springer imprint is published by the registered company Springer Nature Singapore Pte Ltd.
The registered company address is: 152 Beach Road, #21-01/04 Gateway East, Singapore 189721, Singapore

*We would like to dedicate our efforts to our families who inspired and supported us through everything.*

*We would also like to acknowledge support from our students and colleagues for their support throughout.*

# Preface

This book presents the state-of-the-art millimetre wave antennas for next-generation 5G communications. The antenna is an integral part of a communication system, and it is regarded as an intelligent sub-system which can support beamforming, multiple-input, multiple-output (MIMO), massive MIMO and other multiplexing techniques for advanced communication systems. Since propagation losses associated with the millimetre waves (mm-waves) and the signal blockage due to the objects present between transmitter and receiver restrict the performance of the millimetre wave wireless system, novel antenna topologies are required to address such issues.

In the book, various aspects of the antenna design are highlighted related to millimetre wave 5G communication including 28-GHz channel characteristics, mm-wave antenna requirements, antenna design strategies for 28 GHz, MIMO/multibeam antennas and mm-wave lens antennas. Chapter 1 introduces the millimetre wave technology and requirements to implement the mm-wave 5G wireless communication system. The propagation characteristics of the 28-GHz frequency band and different channel measurements are also presented with a brief introduction to beamforming and MIMO. Chapter 2 presents the concept of metamaterials and the realization of anisotropic media using metamaterials with an emphasis on the applications of these engineered materials for antennas. Novel metamaterial-based antennas employing uniaxial/biaxial anisotropic media that enhance the antenna radiation performance are covered in detail. Chapters 3 and 4 introduce some low-profile high-gain antennas with beam tilting capabilities to address high pathloss and misalignment at mm-wave frequencies. The wide-scan and single-/dual-beam MIMO antennas are presented in Chaps. 5 and 6 with emphasis on the diversity performance. The polarization and pattern diversity antennas are presented in Chap. 7. Design and development of dual-beam antennas using metamaterials with an ability to combine the power at receiver side are presented in Chap. 8. Multibeam lens antenna array that results in significant reduction in hardware cost and system complexity for mm-wave massive MIMO applications is also discussed in this book. The design and development of a lens antenna array based on phase gradient metasurface lens are finally presented in Chap. 9 with an introduction to graded refractive index planar lens.

The authors believe that the material presented in this book would be extremely useful to communication and antenna engineers to address the challenges associated with mm-wave 5G wireless communication.

New Delhi, India Shiban Kishen Koul
Jammu, India Zamir Wani

# Contents

# About the Authors

**Shiban Kishen Koul** is Emeritus Professor at the Indian Institute of Technology Delhi. His research interests include RF MEMS, high-frequency wireless communication, microwave engineering, microwave passive and active circuits, device modelling, millimetre and sub-millimetre wave IC design, body area networks, flexible and wearable antennas, medical applications of sub-terahertz waves and reconfigurable microwave circuits including miniaturized antennas. He has successfully completed 38 major sponsored projects, 52 consultancy projects and 61 technology development projects. He has authored/co-authored 523 research papers, 16 state-of-the-art books, 4 chapters and 2 e-books. He holds 23 patents, 6 copyrights and one trademark. He has guided 26 Ph.D. theses and more than 120 master's theses. He is Life Fellow of IEEE, USA, and Fellow of INAE and IETE in India. He is Chief Editor of IETE Journal of Research and Associate Editor of the *International Journal of Microwave and Wireless Technologies*, Cambridge University Press. He served as Distinguished Microwave Lecturer of IEEE MTT-S for the period 2012–2014. He is the recipient of numerous awards including IEEE MTT Society Distinguished Educator Award (2014), Teaching Excellence Award (2012) from IIT Delhi, etc.

**Zamir Wani** received M.Sc. degree in electronics with gold medal from the University of Kashmir, India, in 2012 and the M.Tech. degree in RF and microwave engineering from the Indian Institute of Information Technology, Design and Manufacturing, Jabalpur, India, in 2015. He joined Centre for Applied Research in Electronics, IIT Delhi, as a full-time Ph.D. scholar in January 2016, and completed the degree in August 2020. At present, he is working as Project Scientist at the Indian Institute of Technology Jammu. He has received Raj Mitra Grant in India (RMGI) Award presented at InCAP2018. He has authored or co-authored more than twelve articles in peer-reviewed journals and conference proceedings and has also filed one Indian patent. His main research interests include millimetre wave antennas, MIMO antennas, metamaterials and computational electromagnetics. He serves as Reviewer of IEEE AWPL, IEEE Access, MOTL, IJMWT Cambridge and JEMWA.

# Abbreviations

| | |
|---|---|
| AOA | Angle of Arrival |
| AOD | Angle of Departure |
| AWGN | Additive White Gaussian Noise |
| BF | Beamforming |
| CLL | Capacitively Loaded Loops |
| CST | Computer Simulation Technology |
| DNG | Double Negative |
| ENZ | Epsilon Near Zero |
| FCC | Federal Communications Commission |
| FDTD | Finite Difference Time Domain |
| FEM | Finite Element Method |
| GHz | Gigahertz |
| GRIN | Graded Refractive Index |
| HLE | High-Low Epsilon |
| HPBW | Half-Power Beamwidth |
| ITU | International Telecommunication Union |
| LHCP | Left-Hand Circular Polarization |
| LNA | Low-Noise Amplifier |
| MEMS | Microelectromechanical System |
| MIMO | Multiple-Input, Multiple-Output |
| mm-wave | Millimetre Wave |
| MNZ | Mu Near Zero |
| MPA | Microstrip Patch Antenna |
| MS | Metasurface |
| MTM | Metamaterial |
| NIM | Negative Index Media |
| PDM | Path Division Multiplexing |
| PEC | Perfect Electric Conductor |
| PGMS | Phase Gradient Metasurface |
| PMC | Perfect Magnetic Conductor |
| PRS | Partially Reflecting Surface |

| | |
|---|---|
| RF | Radiofrequency |
| RHCP | Right-Hand Circularly Polarization |
| SM | Spatial Multiplexing |
| SNR | Signal-to-Noise Ratio |
| SPDT | Single Pole Double Throw |
| VNA | Vector Network Analyser |
| VSWR | Voltage Standing Wave Ratio |
| Wi-Fi | Wireless Fidelity |
| WLAN | Wireless Local Area Network |
| WRC | World Radio Conference |

# Chapter 1
# Millimeter Waves for Gigabit Wireless Communication

## 1.1 Introduction

The increasing demand for data rates above 10 Gbit/s is expected to support newer applications like augmented reality, 3D media, 8K video streaming and much more in wireless mobile devices. The demand for high-speed internet and high integrity services will grow in future with a greater number of users adding each day. Existing 3G and 4G services lack the necessary data rates, which motivate one to use millimetre wave (mmWave) spectrum for communication purposes. The mmWave technology is a promising technology for next-generation cellular systems which can offer multi-gigabit per second speeds. In search of higher data rates, the radio systems have moved to operate at higher carrier frequencies to cater to the need. Opening the new frequency bands would help the service providers to accommodate more users while also providing high performance mobile internet to each user. Compared to the frequencies below 6 GHz, mmWave spectrum is lightly licensed and could be made accessible to service providers around the world. The silicon technology has made tremendous progress in manufacturing; the price of mmWave equipment/devices has considerably reduced and it has now become feasible to manufacture consumer electronic products at high frequencies. There are still challenges in implementing mmWave wireless systems which are now being taken up by the researchers.

## 1.2 Millimeter Wave 5G Frequency Band

The large amount of bandwidth available at mmWave frequency bands helps in achieving greater than 10 Gbit/s data rates [1]. For a AWGN channel, the capacity is linearly dependent on the available bandwidth (BW) as:

 
S. K. Koul and Z. Wani, *Novel Millimetre Wave Antennas for MIMO and 5G Applications*, Lecture Notes in Electrical Engineering 819,
https://doi.org/10.1007/978-981-16-7278-1_1

$$C = B_w \log_2\left(1 + \frac{S_i}{N_0 B_w}\right) \tag{1.1}$$

where $S_i$ and $N_0$ are the signal power and noise spectral density [2]. As the signal power is limited due to the regulations in practice, the signal power cannot be increased beyond the safe limit which restricts the mmWave communication to small cell access [3]. The high frequencies of these waves make them useful for transmission of large amount of data in cellular and radar communication. The key benefits of these waves are listed below:

1. High level of frequency re-use enabled: The communication needs of multiple cells within a small geographic region can be satisfied.
2. Fibre optic data transmission speeds possible: This is due to large bandwidth available at these high frequencies.
3. Mature technology: For a very long time these frequencies have been in use for secure communications.

Initial experiments at mmWaves were performed about 130 years ago by Bose and Lebedew [4] and later application of these waves for mobile communications were invented in 1990s. Due to abundant bandwidth available at mmWave spectrum, efforts have been directed largely to develop mmWave wireless networks and systems like IEEE 802.11ad WIFI operating at 60 GHz. Figure 1.1 shows different mmWave

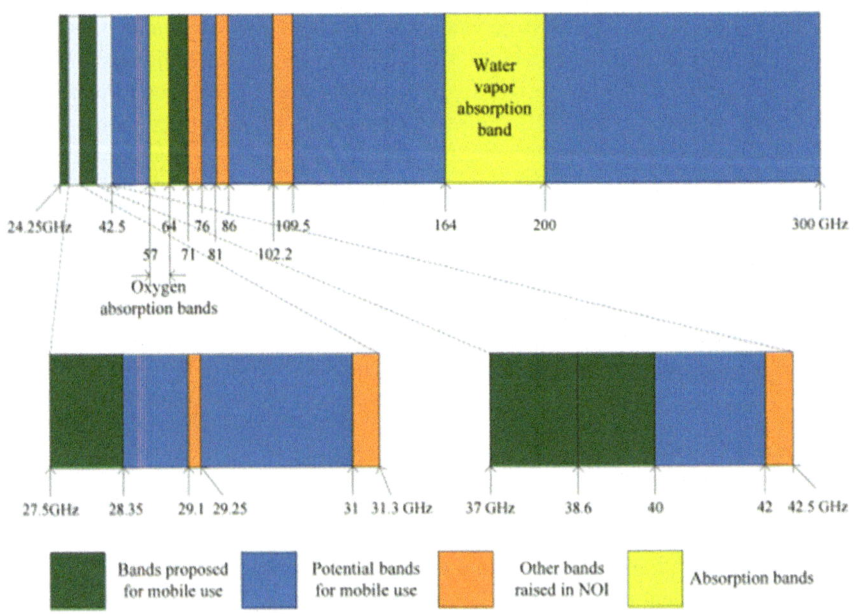


**Fig. 1.1** Millimeter wave spectrum for future mobile communication ([5], Reprinted with permission from IEEE)

frequency bands with potentially large bandwidths available for mobile communications [5]. The available bandwidth exceeds 150 GHz excluding the absorption bands and the unused mmWave bands have already attracted lot of interest. The researchers have investigated some of the best candidate frequencies for use in 5G cellular communications. The 3GPP and the International Telecommunication Union (ITU) have planned two phases of research for 5G standards. The first phase focusing on the use of frequencies below 40 GHz to address more urgent commercial needs and the second phase focusing on frequencies up to 100 GHz. In the World Radio-Communications Conference 2015 (WRC-15) [6], ITU released a list of proposed globally viable frequencies between 24 and 86 GHz. After ITU proposal, the US FCC issued a Notice of Proposed Rule Making (NPRM) on 21st October 2015 that proposed new flexible service rules among the 28 GHz, 37 GHz, 39 GHz, and 64–71 GHz bands [7]. Most recently in WRC-19, additional frequency bands have been identified for next generation 5G deployment which include 24.5–27.5 GHz, 37–43.5 GHz, 45.5–47 GHz, 47.2–48.2 GHz and 66–71 GHz bands.

The three frequency bands (28 GHz, 39 GHz and 73 GHz) have emerged as potential candidates for long distance communication due to their low oxygen absorption unlike absorption band centred at 60 GHz [8].

### 1.2.1 The 28-GHz Band

The use of 28-GHz frequency band for the wireless communication requires a-priori knowledge of the losses associated with the propagation of these waves as they travel through the atmosphere. The losses associated with the propagation of the waves at this band are free space and atmospheric loss. The free space loss is due to spherical spreading of the waves as they propagate, the power falls as the square of the propagation distance which equivalently means the received power depends on $(1/r^2)$. For a distance of 100 m, the free space path loss for existing 5 GHz system and 28 GHz system is 86.42 dB and 101.4 dB respectively, which means the loss difference is about 15 dB. Now in order to cover this loss at the 28 GHz, high gain antennas are essential. Apart from this loss, the other atmospheric losses are due to oxygen, water droplets and other gasses present in the atmosphere. The absorption of the mmWaves has been studied and it is observed that and at certain frequencies the losses are more prone. The average attenuation due to water vapor and oxygen for the 28 GHz and 60 GHz is about 0.15 dB/km and 15 dB/km as they propagate through the atmosphere which translates to 0.03 dB loss for 200 m distance for a 28 GHz system and is negligibly small for a 200 m cell size [9].

Considering the low oxygen and water vapor absorption at 28 GHz, the 28-GHz band has emerged as one of the potential carrier frequency for 5G (fifth-generation wireless communication) to address the growing demand for high speed internet [10]. Samsung Corporation have already initiated work on the use of mmWave frequencies since February 2015. They have performed the channel measurements at 28 GHz to show the viability of this frequency band for cellular communications.

By performing the measurements, Samsung claims that mmWave communication is possible over a 200 m distance [10, 11]. In Japan, there was a collaboration among various telecommunications companies to explore and run field trials at 28 GHz.

### 1.2.2 Propagation Losses

As the electromagnetic waves travel though the free space, the power falls off as square of the distance. Thus, if mmWaves travel double the distance, power reaching the receiver is reduced by a factor of four. This is the result of the spherical spreading of electromagnetic waves as they propagate. The loss factor between transmitter and receiver antenna can be expressed in terms of distance and operating frequency by the following relation:

$$L_{free\,space}(\mathrm{dB}) = 20\log_{10}\left(4\pi\frac{d}{\lambda}\right) \tag{1.2}$$

where $L_{free\,space}$ is the free space loss, $d$ is the distance between transmitter and receiver antennas, and $\lambda$ is operating wavelength. This equation shows that free space loss increases as the operating frequency increases. For millimetre waves, free space loss is quite high even for short distances. This equation suggests that the mmWave spectrum is best for short distance communications.

For a distance of $d = 10\,\mathrm{m}$, the path loss can be calculated. Table 1.1 shows calculated path loss for different frequency bands.

The path loss difference between 28 GHz and existing 5 GHz communication band is about 15 dB, this loss can be compensated for by the use of high gain antenna and choosing suitable architecture.

More generally, the power received by an antenna can be expressed by the *Friis formula*, it gives a more complete picture accounting for all factors [12]

$$P_R(d) = P_T G_T G_R \frac{\lambda^2}{(4\Pi)^2 d^n} \tag{1.3}$$

where $P_R$ is the received power, $P_T$ is the transmitted power, $G_T$ is the transmitting antenna gain, $G_R$ is the receiving antenna gain, $d$ is the distance between the receiver

**Table 1.1** Path loss at different frequency bands

| Frequency band (GHz) | Path loss (dB) |
|---|---|
| 2.4 | 60 |
| 5 | 66 |
| 28 | 81 |
| 39 | 84 |
| 60 | 88 |

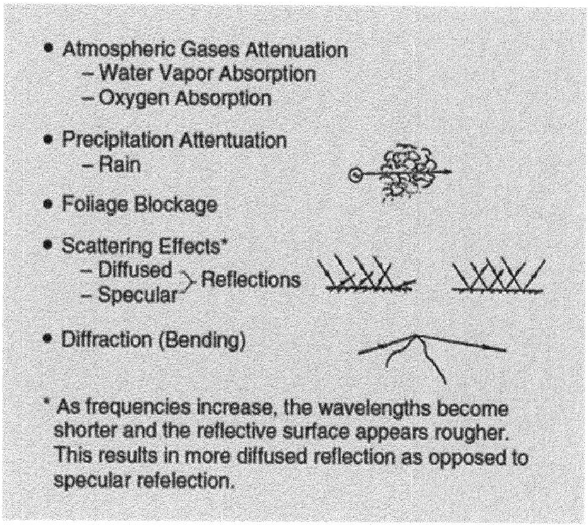


**Fig. 1.2** Propagation effects influencing mmWave propagation ([9], Reprinted with permission from IEEE)

and the transmitter, $n$ is the path loss exponent which is 2 for free space and $\lambda$ is the operating wavelength. For non-free space propagation, (1.3) can be used to approximately find the received power with a suitable value of $n$ found using channel measurements.

The main source of transmission loss is free space loss given by (1.2). Apart from this loss, there are other loss factors, such as losses due to gases and losses due to rain present in the transmission medium. The millimeter waves are affected by the factors which are shown in Fig. 1.2. The loss in the atmosphere is due the absorption of mmWaves by oxygen and water vapor and other gaseous constituents present, as these waves travel through atmosphere. Figure 1.3 shows the attenuation of the waves in the atmosphere, the losses are prone to be more at certain frequencies coinciding with natural resonant frequency of the gas molecules.

### *1.2.3 Channel Measurements*

Researchers have already started working on developing channel models for indoor and outdoor environments at 28 GHz, 39 GHz and 73 GHz. Dr. Ted Rappaport and his students at NYU Polytechnic School of Engineering in USA have published several papers based on measurements carried out on propagation of 28 GHz frequencies. In [13], authors have performed propagation measurements in typical indoor office environment at 28 GHz using two pairs of directional antennas with 15 and 20 dBi gain separated by a distance ranging from 3.9 to 45.9 m. The antennas are rotated in both the planes to collect the angle of arrival (AOA) and angle of departure (AOD) statistics in indoor environments. The measurements show a path loss exponent of 1.1 for a 1 m distance in close line of sight measurements, which is better than free

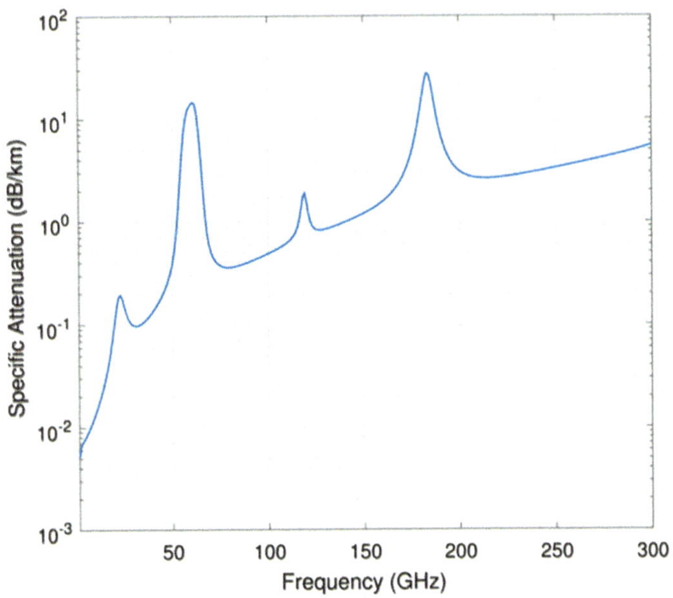


**Fig. 1.3** Attenuation of the signal for different frequency bands ([5], Reprinted with permission from IEEE)

space propagation. The reason being constructive interference of multipath signals. The directional path loss exponent of 4.1 is reported in the paper at 28 GHz, while considering unique pointing angles, this was decreased to 3.0 when receiver antenna was pointed towards strong signal angle.

Figure 1.4 shows the maximum coverage distance as a function of combined transmitter–receiver antenna gain for various values of exponent $n$ for the mmWave system operating at 28 GHz. The plot shows signal coverage for different environments characterized by the different n values (higher value of n signifies highly obstructed environment, $n = 2$ for free space). We can observe from blue curves in the plot that with 30 dBi combined gain (15 dBi at Tx and 15 dBi at Rx), a 350 m propagation distance can be covered in obstructive environment having $n = 4$. Further, as observed from the plot, the coverage range can be increased by increasing the antenna gains.

### *1.2.4 Link Budget*

Link budget is used to determine the usability of RF transceivers. Noise figure of transmitter and receiver is used as a parameter to determine the coverage range under several operating conditions [8]. In the following, calculations representing the estimate of the acceptable communication link range at 28 GHz are presented.

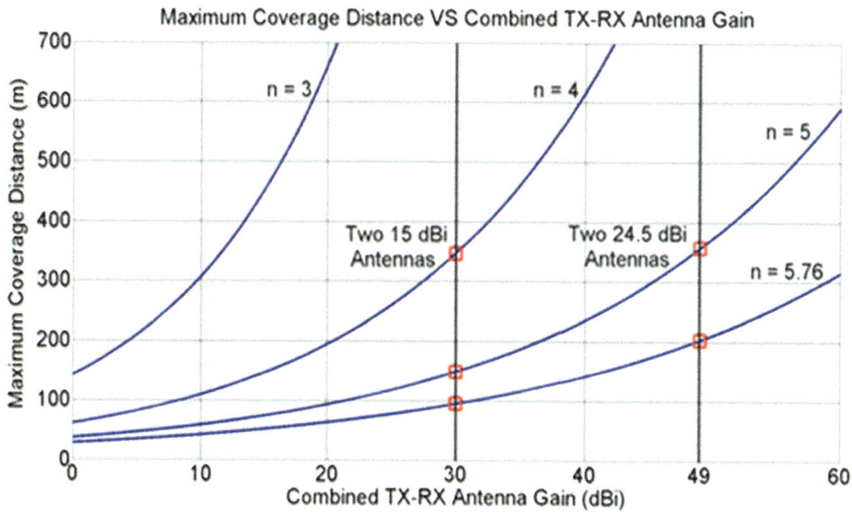


**Fig. 1.4** Maximum coverage with 10 dB SNR versus combined Tx-Rx antenna gain for different path loss exponent $n$ ([10], Reproduced with permission of IEEE)

Link budget with measured parameters is presented in [13] at 28 GHz resulting in a 2.5 Gbps data rate. The signal power received ($P_r$ in dBm) is given below [14]:

$$P_r = P_t - P_l + G_t + G_r \tag{1.4}$$

where $P_t$ = transmitted power (dBm), $P_l$ = path loss, $G_t$ = transmitting antenna gain and $G_r$ = receiving antenna gain.

For 28 GHz, the free space path loss at 1 m is

$$P_l = 61.38\,\text{dB}.$$

The noise power in dBm at the input of the receiver is given by [15]

$$P_n = 10\log(kT) + 10\log(BW) + NF_r \tag{1.5}$$

where $NF_r$ is the noise figure of the receiver in dB, BW is the bandwidth available at 28 GHz. Also $10\log kT = -174\,\text{dBm/Hz}$ for 17 °C system temperature. The received power from (1.4) is given below for 10 dBm input transmitter power.

$$P_r = 10\,\text{dBm} - 61.38\,\text{dB} + 0 + 0 = -51.38\,\text{dBm}$$

Let us assume LNA noise figure is 6 dB and bandwidth at 28 GHz is 0.85 GHz, then the noise power at the input of the receiver is given by (1.5) as

$$P_n = -\frac{174\,\text{dBm}}{\text{Hz}} + 10\log(0.85\,\text{GHz}) + 6\,\text{dB} = -78.7\,\text{dBm}$$

The signal to noise ratio (SNR) for the 1 m link is given below

$$SNR(\text{dB}) = signal\ power - noise\ power = -51.38 - (-78.7) = 27.3\,\text{dB}$$

This SNR is sufficient to use 64-QAM modulation (which requires 21.7 dB SNR) scheme to achieve giga-bit/s data rates. The above calculations are for 1 m link with unity gain antennas at both ends. The same performance can be achieved even for larger distances by using high gain antennas.

## 1.3 Millimeter Wave Antennas

Antenna is the most important element of a communication system. It acts as the transducer to couple the signals into the communication channel and can sense the incoming electromagnetic signals for detection. All the communicating devices such as mobile phones, wireless routers etc. have built-in-antennas to transmit and receive signals, but nowadays hardly one can see these antennas. The size of the antenna is mainly decided by the frequency which it is intended to operate upon, higher the frequency smaller is the antenna size. The demand of high data rate in communication increases day by day. In order to cater to the needs of present-day, wireless communication networks are shifting to high carrier frequencies, thus decreasing the size of antennas. Today's devices use advanced microstrip antennas for communications. The present day antennas use a thin metal strip mounted on a dielectric substrate that is attached to the outer rim of the device. This approach effectively hides the antenna under the mobile device outer cover.

The high path loss seems to be a problem at mmWave frequencies, this will restrict the communication at these frequencies to indoor environments only. In addition, a high frequency reuse can also be achieved over a short distance, which allows a higher throughput network. The compact size of antenna at these frequencies allows us to design multiple antennas in a small space available, thus making communication systems compact. The typical antennas that are used for mmWave communications are reflector, lens and horn antennas [16]. These types of antennas provide high gain, however they are not preferred for commercial handheld devices because of their large size and higher cost. The simplest and easy to fabricate are printed circuit antennas that can be fabricated using photo-lithography process. The microstrip patch or dipole array antennas are low profile, light weight and can be used in small handheld planar devices.

### 1.3.1 Antenna Design and Requirements

The use of 28-GHz frequency for next-generation 5G communication purpose offers far higher channel capacity than use of lower frequencies. One of the key issues at the mmWave frequency is high propagation losses which require high gain antennas for better SNR. The narrow beamwidth associated with highly directive antennas makes the beam alignment very difficult. A mmWave antenna with a gain of 20 dBi has roughly 16° of half power beamwidth (HPBW). In order to maintain a reliable communication link with this type of antenna at 28-GHz is quite difficult due to the misalignment of either transmitter or receiver antenna. In order to solve this issue, a wide-scan or switched beam antenna array can be implemented to switch the radiation beams to search for the available signal.

While modelling of the communication channel in between the transmitter and the receiver antennas, the directivity is usually considered constant with large value for a narrow angular range and zero elsewhere [17]. However, at mmWave frequency band, practical high directional antennas have more complicated radiation patterns with sharp main lobe radiation and non-zero sidelobe levels. The 3D radiation pattern of a typical microstrip patch antenna array is shown in Fig. 1.5, the practical 4 × 4 rectangular patch antenna array with corporate feed has 15.5 dBi of gain and has −2.4 dB and −13.6 dB of side lobe levels in the E-plane and H-plane, respectively with respect to the main beam. Therefore, the effect of reduced main lobe gain due

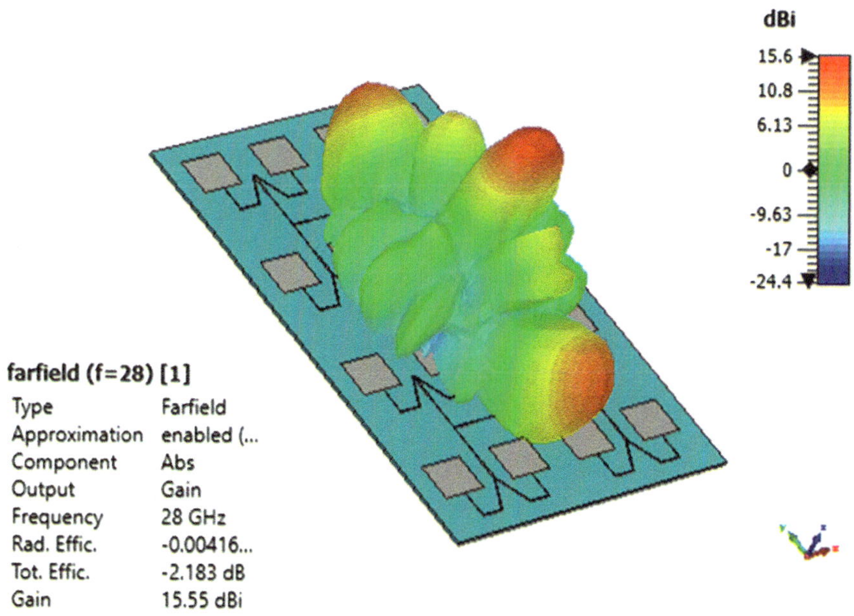


**Fig. 1.5** 3D radiation pattern of a typical 4 × 4 microstrip patch antenna array

to misalignment and side lobe radiation should be taken into consideration while design of a mmWave wireless system.

## 1.4 Millimeter Wave MIMO

Multiple-Input-Multiple-Output (MIMO) system improves the communication performance by utilizing multiple antennas at both the transmitter and receiver ends. The communication range and throughput of the communication system is increased significantly using MIMO. The MIMO technique is used in 4G LTE cellular communication and is planned for mmWave communications as well to improve the link reliability and capacity [18]. Most common approaches to use MIMO systems to full potential are spatial multiplexing (SM) and beamforming (BF). In [18], authors have described how SM and BF can best be used for mmWave wireless communication systems. The factors that define when to use SM and when BF are signal-to-noise ratio (SNR) and bandwidth provided to each user. In SM, data is divided into multiple streams and sent on a single carrier to increase the capacity per user, this is effective when communications links are having high SNR and are limited in bandwidth but not power limited. At low SNR (power limited channels), SM divides power into multiple streams thus weakens each stream and chances of some streams might be lost, introducing bit errors. Therefore, SM provides little benefit at low SNR while as BF can provide increased capacity by increasing SNR. The mmWave systems are power limited rather than bandwidth limited. The mmWave devices will accommodate dozens of antennas (due to small wavelengths) to provide more dimensions in the communication channel, together SM and BF can be used to increase channel throughput [19, 20]. The implementation of the MIMO system can be done using three different architectures such as fully digital, analog and hybrid architecture as shown in Fig. 1.6. The fully digital MIMO architecture shown in Fig. 1.6a requires number of RF chains equal to the number of antennas, which leads to high energy consumption in RF chains and un-affordable hardware cost. The analog architecture implementation has less complexity and requires a single RF chain as shown in Fig. 1.6b. A single data stream is transmitted with an array gain realized using the phase shifting network essential to compensate for transmission losses at mmWaves. However, with this architecture spatial multiplexing cannot be achieved as only one RF chain is present [5, 18]. The BF and SM can both be implemented using the hybrid MIMO architecture (as in Fig. 1.6c) and is the key solution for mmWave systems [21]. This kind of architecture can be considered as the extension of the analog MIMO architecture with fewer RF chains which reduces the power consumption and cost.

In Sect. 1.2, we have discussed the propagation losses associated with mmWave communication and the solution using high gain antennas. However, the shadowing issue of the mmWave signals due to obstacles and human body needs considerable attention. To overcome this issue of shadowing in mmWave communication, MIMO technique can be utilized to build multiple channels between the transmitter and receiver, thereby choosing the best received signal for maintaining the link. The

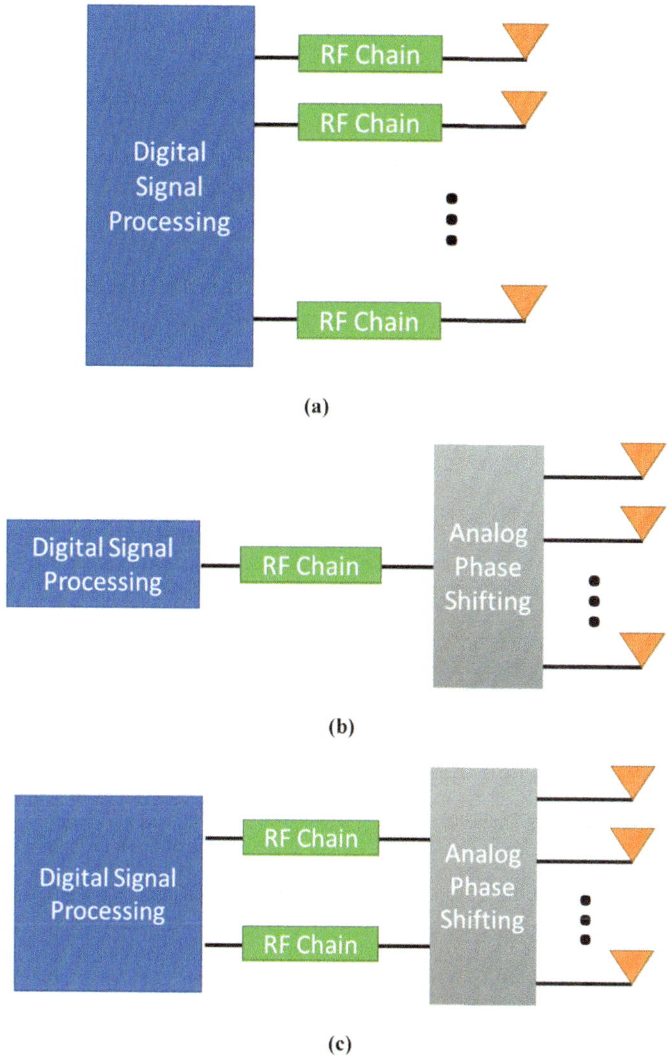


**Fig. 1.6** Possible MIMO architecture for mmWave systems, **a** fully digital, **b** analog and **c** hybrid architecture

mmWave communication indoors is mainly non-LOS scenario, the issue of shadowing can be solved using MIMO antennas with multiple radiation beams to reduce loss from obstacles.

## 1.5 Conclusion

This Chapter introduces the millimetre wave technology and requirements to implement the mmWave 5G wireless communication system. Number of studies describing the use of 28-GHz frequency band for 5G communication are presented. The propagation characteristics of the 28-GHz frequency band and different channel measurements are also presented. Also, a link budget study is presented to show the performance of the system at 28-GHz. Different antenna topologies for the mmWave frequency band and various design requirements are briefly discussed. Finally, the advantages of MIMO and beamforming technique for the mmWave wireless communication are presented.

## References

1. Pi Z, Khan F (2011) An introduction to millimeter-wave mobile broadband systems. IEEE Commun Mag 49(6):101–107. https://doi.org/10.1109/MCOM.2011.5783993
2. Shannon CE (1948) A mathematical theory of communication. Bell Syst Tech J 27(3):379–423. https://doi.org/10.1002/j.1538-7305.1948.tb01338.x
3. Ge X, Cheng H, Guizani M, Han T (2014) 5G wireless backhaul networks: challenges and research advances. IEEE Netw 28(6):6–11. https://doi.org/10.1109/MNET.2014.6963798
4. Emerson DT (1997) The work of Jagadis Chandra Bose: 100 years of millimeter-wave research. IEEE Trans Microw Theory Tech 45(12):2267–2273. https://doi.org/10.1109/22.643830
5. Xiao M et al (2017) Millimeter wave communications for future mobile networks. IEEE J Sel Areas Commun 35(9):1909–1935. https://doi.org/10.1109/JSAC.2017.2719924
6. Resolution Com6/20, Provisional final acts WRC-15. WRC-15. ITU, Geneva, pp 424–426. https://www.itu.int/dms_pub/itu-r/opb/act/R-ACT-WRC.11-2015-PDF-E.pdf
7. Use of spectrum bands above 24 GHz for mobile radio services, GN Docket No. 14-177, Notice of Proposed Rulemaking, 15 FCC Record 138A1 (rel. Oct. 23, 2015)
8. Rappaport TS, Murdock JN, Gutierrez F (2011) State of the art in 60-GHz integrated circuits and systems for wireless communications. Proc IEEE 99(8):1390–1436. https://doi.org/10.1109/JPROC.2011.2143650
9. Marcus M, Pattan B (2005) Millimeter wave propagation; spectrum management implications. IEEE Microw Mag 6(2):54–62. https://doi.org/10.1109/MMW.2005.1491267
10. Rappaport TS et al (2013) Millimeter wave mobile communications for 5G cellular: it will work! IEEE Access 1:335–349. https://doi.org/10.1109/ACCESS.2013.2260813
11. Samsung '5G Vision', p 7. http://www.samsung.com/global/business-images/insights/2015/Samsung-5G-Vision-0.pdf
12. Friis HT (1946) A note on a simple transmission formula. Proc IRE 34(5):254–256. https://doi.org/10.1109/JRPROC.1946.234568
13. Deng S, Samimi MK, Rappaport TS (2015) 28 GHz and 73 GHz millimeter-wave indoor propagation measurements and path loss models. ArXiv150900395 Cs Math, pp 1244–1250. http://doi.org/10.1109/ICCW.2015.7247348
14. Rappaport TS (2002) Wireless communications: principles and practice, 2nd edn. Prentice Hall, Upper Saddle River, N.J
15. Leinonen ME, Destino G, Kursu O, Sonkki M, Pärssinen A (2018) 28 GHz wireless backhaul transceiver characterization and radio link budget. ETRI J 40(1):89–100. https://doi.org/10.4218/etrij.2017-0231
16. Dybdal R (1983) Millimeter wave antenna technology. IEEE J Sel Areas Commun 1(4):633–644. https://doi.org/10.1109/JSAC.1983.1145968

17. Haneda K et al (2016) Indoor 5G 3GPP-like channel models for office and shopping mall environments. In: 2016 IEEE international conference on communications workshops (ICC), pp 694–699. http://doi.org/10.1109/ICCW.2016.7503868
18. Sun S, Rappaport TS, Heath RW, Nix A, Rangan S (2014) Mimo for millimeter-wave wireless communications: beamforming, spatial multiplexing, or both? IEEE Commun Mag 52(12):110–121. https://doi.org/10.1109/MCOM.2014.6979962
19. Lee SJ, Lee WY (2013) Capacity of multiple beamformed spatial stream transmission in millimetre-wave communication channels. IET Commun 7(12):1263–1268. https://doi.org/10.1049/iet-com.2012.0747
20. Hong W, Baek K, Lee Y, Kim YG (2014) Design and analysis of a low-profile 28 GHz beam steering antenna solution for future 5G cellular applications. In: 2014 IEEE MTT-S international microwave symposium (IMS2014), pp 1–4. http://doi.org/10.1109/MWSYM.2014.6848377
21. Zhang X, Molisch AF, Kung S-Y (2005) Variable-phase-shift-based RF-baseband codesign for MIMO antenna selection. IEEE Trans Signal Process 53(11):4091–4103. http://doi.org/10.1109/TSP.2005.857024

# Chapter 2
# Metamaterials and Anisotropic Media Realization for mmWave Applications

## 2.1 Introduction

Metamaterials are engineered materials which gain their properties depending on the macroscopic composite of periodic/non-periodic structure. According to Rodger M Walser, metamaterials are defined as "Macroscopic composites having a synthetic, three-dimensional, periodic cellular architecture designed to produce an optimized combination, not available in nature, of two or more responses to specific excitation" [1]. These designed materials contain array of small "model atoms" to generate specific magnetic or dielectric response for an electromagnetic wave. The model atoms are basically subwavelength resonators designed using metallic structures providing resistive-inductive-capacitive response to mimic the electric or magnetic response in bulk materials. Based on the electric and magnetic field interaction in the materials, different materials can be characterized by electric permittivity ($\epsilon$) and magnetic permeability ($\mu$). The refractive index of a material is another important parameter and is defined as $n = \sqrt{\epsilon_r \mu_r}$, where $\epsilon_r = \epsilon/\epsilon_0$ and $\mu_r = \mu/\mu_0$ is ratio of permittivity/permeability of material to permittivity/permeability of free space. Based on the values of $\epsilon$ and $\mu$, different materials can be characterized as shown in Fig. 2.1. This figure depicts different materials with different properties based on the values of $\epsilon$ and $\mu$. The first quadrant in the figure corresponds to the materials having $\epsilon > 0$ and $\mu > 0$, these materials support forward travelling waves and are known as right-handed materials (RHM). Naturally occurring RH materials has $\epsilon \geq \epsilon_0$ and $\mu = \mu_0$ and depicted along a line in the first quadrant. The second quadrant in the figure has $\epsilon < 0$ and $\mu > 0$ which represents electric plasma and can support evanescent waves. The materials having $\epsilon < 0$ and $\mu < 0$ depicted in third quadrant were proposed by Veselago [2] are called left-handed materials (LHM). In LHM, the k-vector forms left hand system with electric filed and magnetic field in the materials, thus supports backward wave propagation. The materials in the fourth quadrant have $\epsilon > 0$ and $\mu < 0$ are of magnetic plasma type and support evanescent wave.

 
S. K. Koul and Z. Wani, *Novel Millimetre Wave Antennas for MIMO and 5G Applications*, Lecture Notes in Electrical Engineering 819,
https://doi.org/10.1007/978-981-16-7278-1_2

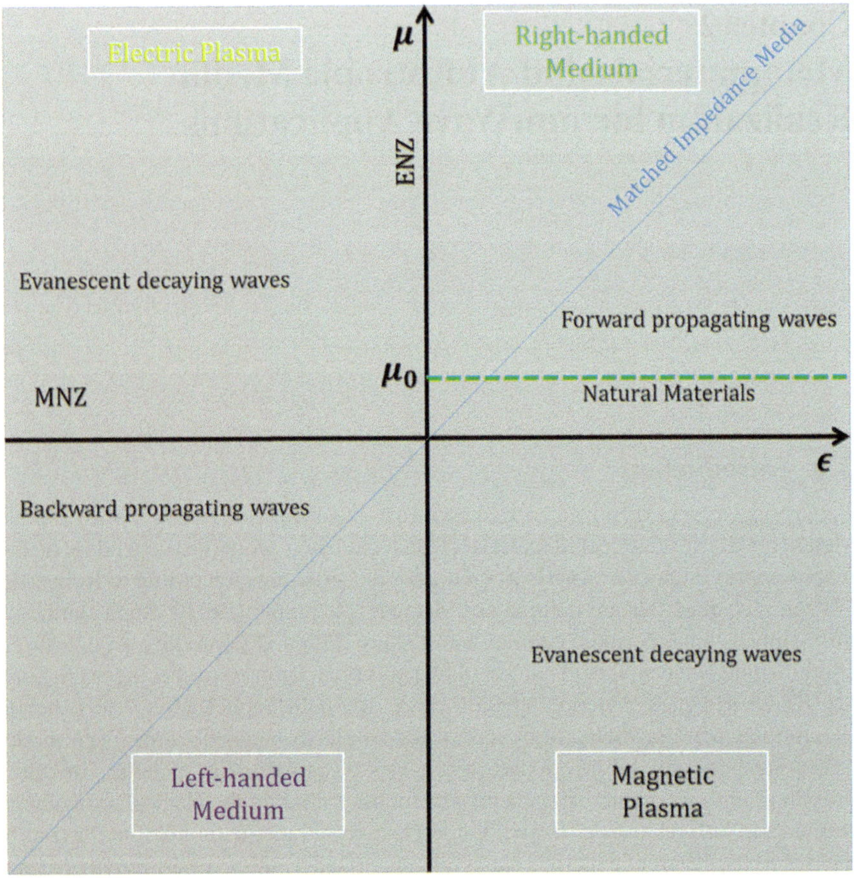


**Fig. 2.1** Classification of materials in $\epsilon - \mu$ domain

The naturally occurring materials lying in the first quadrant are nonmagnetic materials with epsilon greater than the free space and some lie in second and fourth quadrants which are of electric plasma and magnetic plasma type. The other materials in quadrant first (RHM) and third (LHM) can be realized using metamaterials by using different cellular architecture and arrangements. Other special materials such as epsilon near zero (ENZ) having $\epsilon = 0$, mu near zero (MNZ) having $\mu = 0$ are depicted in Fig. 2.1. The materials in the third quadrant having $\epsilon = -\epsilon_0$ and $\mu = -\mu_0$ has very special property, these materials are perfect match to free space and can be used for designing a perfect lens. Apart from the materials depicted in Fig. 2.1, different anisotropic medias can also be designed using metamaterials. The flexibility to realize different materials using metamaterials helps in controlling the electromagnetic propagation through these materials at will.

## 2.2 Metamaterials for Antennas

The metamaterials have found several novel applications since the discovery of negative index metamaterials (NIM) designed using SRRs and wire structures fabricated on PCB sheets by Dr. Smith in 2001 [3]. Several years later, with further research in the area of metamaterials, bending of electromagnetic waves using gradient index metamaterials, near zero index materials and the invisible cloaks became possible. Metamaterials due to the extraordinary features and realization of customized material properties have found potential applications especially in antenna engineering. The applications of metamaterials for antennas include bandwidth enhancement, miniaturization, gain enhancement, controlling the beam direction, and polarization change. Some of the applications of metamaterials to achieve desired antenna performance are discussed below.

### 2.2.1 *Antenna with NIM*

The antennas with reconfigurable radiation pattern are quite useful for wireless communications systems to solve interference and blockage problems. Integrated beam tilting antennas having certain tilt in the radiation directions resulting in decreasing the interface for that direction which enhances the system throughput. Conventionally beam tilting is done electronically or mechanically, however these two techniques suffer with system losses and require mechanical installation framework.

One of the practical ways of achieving beam tilt is done using negative index metamaterials as shown in [4]. The antenna consists of two dielectric resonator antennas and in order to tilt the radiation beam directions, a negative index metamaterial is placed in between the two antennas. With the help of negative index metamaterial, the electromagnetic waves from the antennas are refracted which result in change in the direction of the radiation as shown in Fig. 2.2. The negative index metamaterial

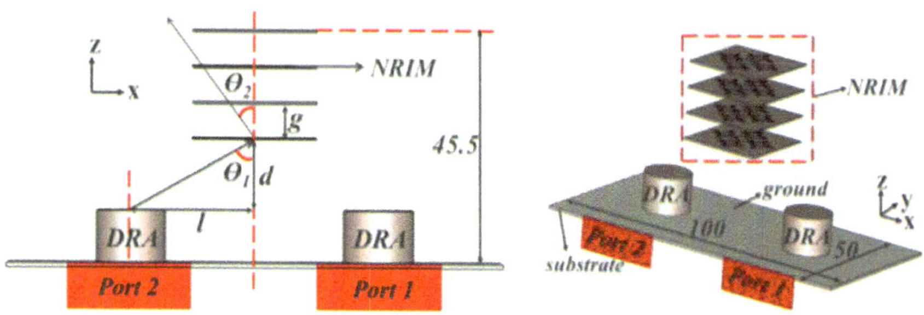


**Fig. 2.2** Geometry of the two-antenna systems with negative index metamaterials ($l$ = 25, $d$ = 14, $g$ = 7, all in mm) ([4], Reprinted with permission from IEEE)

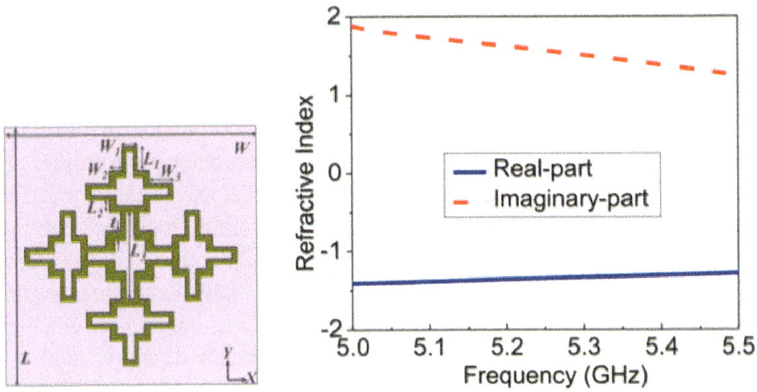


**Fig. 2.3** Unit cell design and extracted refractive index ($W = L = 25$, $W_1 = 1.5$, $L_1 = W_3 = 2.35$, $W_2 = L_2 = 1.25$, $L_3 = 8.4$, $t = 0.9$, all in mm) ([4], Reprinted with permission from IEEE)

is realized using the fractal cross-ring unit cells to achieve negative refractive index for the 5–5.5 GHz band. The designed unit cell and the extracted parameters for the electromagnetic excitation along z-axis are shown in Fig. 2.3. It is clear that the unit cell provides negative refractive index from 5 to 5.5 GHz.

The negative index metamaterial consists of $1 \times 4$ unit cell layers printed on RT-duroid 5880 substrate with height h = 0.254 mm. The radiation performance of the antenna with NIM results in a 38° of beam tilt as shown in the radiation pattern of the antenna when port 1 is active as shown in Fig. 2.4a. The radiated fields shown in the Fig. 2.4b represent the negative refraction due to the NIM in between two ports. The beam tilt is given by the Snell's law ($n_1 \sin\theta_1 = n_{NIM} \sin\theta_2$) by equating the tangential k-vectors at the boundary of NIM and air. Since the $n_{NIM}$ is negative, the refraction angle ($\theta_2$) at NIM and air is about 38.5° from Fig. 2.2. The measured radiation performance shows the beam tilting in xoz-plane with the radiation beam along −38° and +38° over the frequency band.

### 2.2.2 *Antenna with ZIM*

The zero index metamaterials (ZIM) with permittivity or permeability equal or close to zero has very special applications in antennas. The ZIMs are utilized to enhance the directivity of the antennas when placed atop. The ZIM is realized using subwavelength metallic wires in the microwave and visible region. The gain or directivity of the antennas is enhanced using the ZIMs realized using the multi-layered substrates as demonstrated in [5, 6].

A high gain circularly polarized antenna with zero index metamaterial is also proposed in [7]. Here the zero-index metamaterial consists of array of square ring with two asymmetrical splits printed on FR4 substrate (thickness = 0.8 mm) as

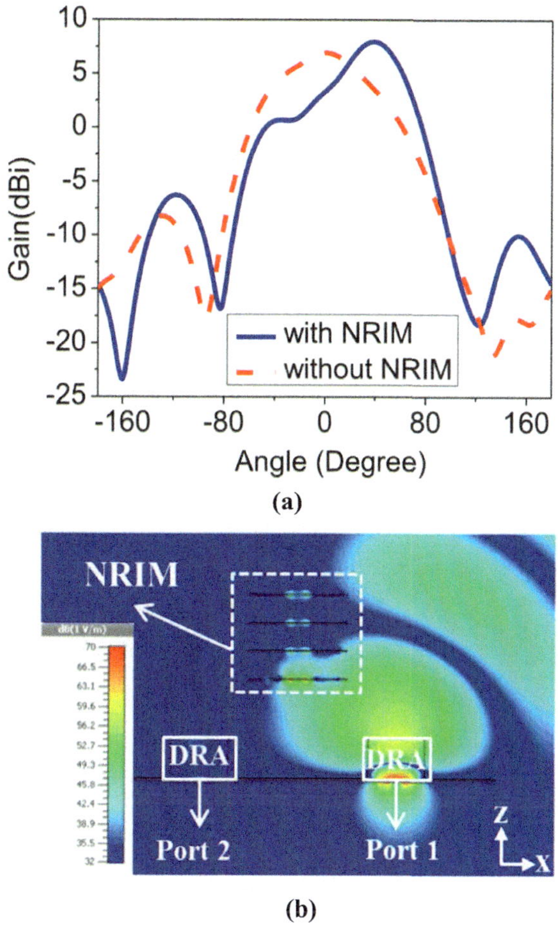


**Fig. 2.4** **a** Radiation pattern when port 1 is active and **b** electric field distribution in xoz-plane ([4], Reprinted with permission from IEEE)

shown in Fig. 2.5a. Each unit cell has two identical square rings etched back-to-back on the substrate with split gaps to make it work with two orthogonal polarizations. The unit cell's electromagnetic response is studied using full wave simulations and the constituent material parameters are plotted in Fig. 2.5b. From the figures, it is clear that the unit cell has a zero refractive index at 7.45 GHz.

The zero index metamaterials consisting of $9 \times 9$ unit cells are placed at a height of 23.3 mm from the radiating patch antenna at 7.45 GHz as shown in Fig. 2.6. As per the theory of ZIM, the fields incident on ZIM from the patch antenna would be refracted towards the broadside direction which results in the enhanced gain of the antenna. The gain of the patch antenna with and without ZIM is plotted in Fig. 2.7, the achieved gain enhancement is about 5–6 dB. The antenna with ZIM has circularly

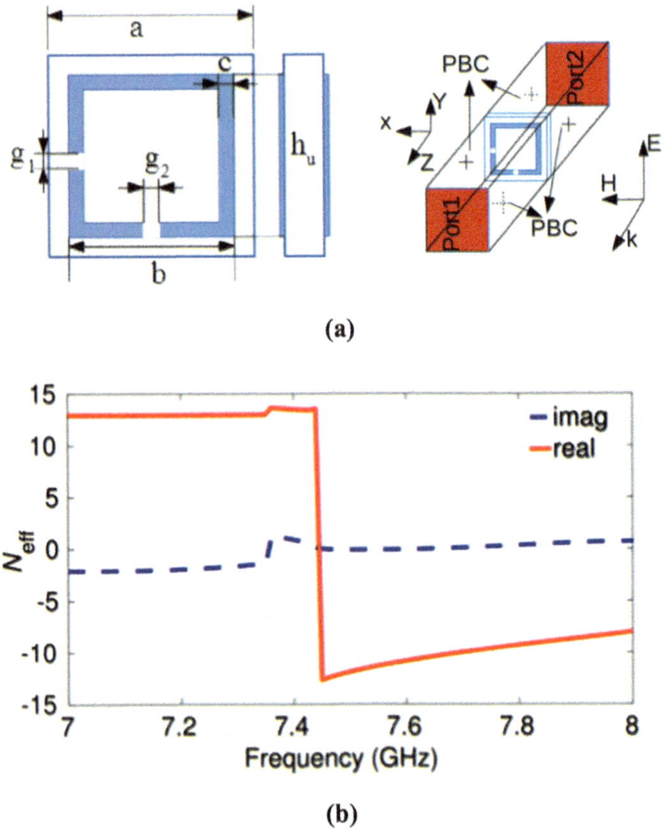


**Fig. 2.5** **a** Unit cell geometry and simulation setup and **b** extracted effective refractive index versus frequency ([7], Reprinted with permission from IEEE)

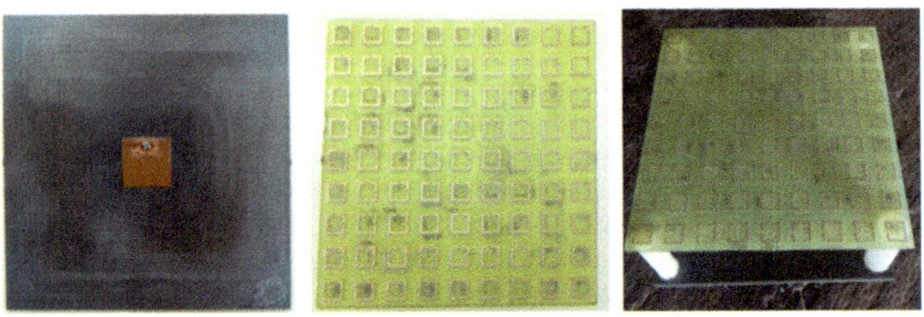

**Fig. 2.6** Fabricated patch antenna with ZIM superstrate ([7], Reprinted with permission from IEEE)

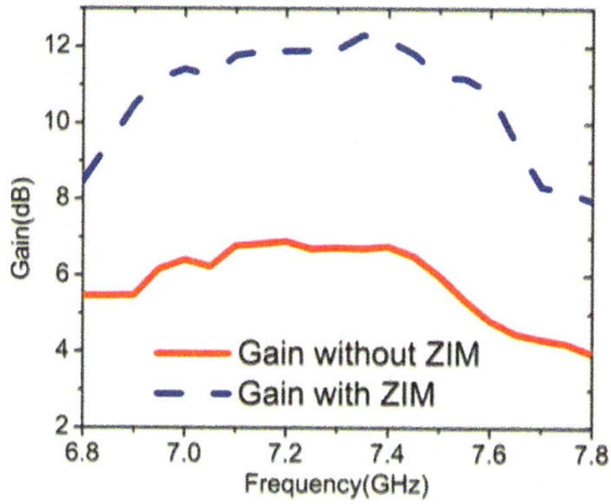


**Fig. 2.7** Gain comparison with and without zero index metamaterial ([7], Reprinted with permission from IEEE)

polarized (CP) radiation with the fussing of the fields at 7.45 GHz, resulting in a measured peak gain of 12.3 dBi.

## 2.3 Anisotropic Media and Applications

The anisotropic media with extreme parameters, such as high permittivity or permeability along one of the axes offers efficient transport of electromagnetic energy. These materials find applications in photonic waveguides and lenses due to very large index [8, 9]. In case of infinite anisotropy (meaning infinite permittivity along one direction), diffraction-free electromagnetic wave routing can be achieved with no reflected waves [10]. The property of diffraction free transfer of EM waves in high epsilon anisotropic medias can be studied from the dispersion relations in the media. The following section describes the anisotropic media properties along with the realization of the media using metamaterials.

### *2.3.1 Media Properties*

The uniaxial anisotropic medium (with one component of permittivity different than other two) can be defined by the permittivity tensor ($\overline{\overline{\epsilon}}$). We consider a uniaxial medium with permeability $\mu_0$ and permittivity tensor given by:

$$\overline{\overline{\epsilon}} = \epsilon_0 \begin{bmatrix} \epsilon_x & 0 & 0 \\ 0 & \epsilon_\perp & 0 \\ 0 & 0 & \epsilon_\perp \end{bmatrix} \tag{2.1}$$

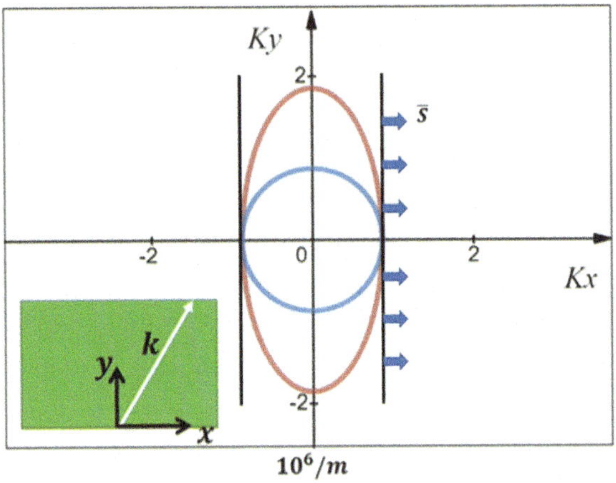


**Fig. 2.8** Iso-frequency contour for TE waves (blue circle) and TM waves for $\epsilon_x = 10$ (red) and for infinite anisotropy $\epsilon_x = \infty$ (black lines) and $\epsilon_\perp = 2.2$

where $\epsilon_0$, $\epsilon_x$ and $\epsilon_\perp$ are the free space permittivity, relative permittivity along optical axis (x-axis) of the medium and relative permittivity transverse to optical axis respectively. Let us consider a 2D plane z = 0 in the anisotropic media with epsilon vector $\overline{\overline{\epsilon}}$; in which the field configuration TM $(Ex, Ey, Hz)$ and the propagation is in xy-plane. The propagation vector, $k = k_x\overline{x} + k_y\overline{y}$ encounters different refractive index in different direction in this plane so the waves are referred to as extraordinary waves. While in the same plane with field configuration TE (Hx, Hy, Ez), these waves would propagate with same speed along all the directions, and hence these are ordinary waves. The iso-frequency contour for the TM and TE fields in xy-plane is shown in Fig. 2.8. Let's consider a dipole antenna as a source, the fields from a dipole antenna can be considered superposition of transverse electric (TE) and transverse magnetic (TM) waves [11]. For a dipole placed in xy-plane with its dipole moment along x-axis, most of the fields in air are TM polarized in xy-plane. The TM fields from the dipole entering the uniaxial media which has high epsilon along the x-axis are having E-field in xy-plane and H-field along the z-direction, these waves would encounter different refractive indices (between $\sqrt{\epsilon_\perp}$ to $\sqrt{\epsilon_x}$) based on the propagation direction (extraordinary waves). The propagation of extraordinary waves in this media is governed by dispersion relation given as [12]:

$$\frac{k_x^2}{\epsilon_\perp} + \frac{k_\perp^2}{\epsilon_x} = k_0^2 \tag{2.2}$$

where $k_x$ and $k_\perp$ are the propagation vectors along optical axis (x-axis) and transverse axis (y-axis) of the medium and $k_o = \omega\sqrt{\mu_o\epsilon_o}$ is the propagation constant in free space. The iso-frequency contour of the extraordinary waves is an ellipse with radius along optical axis being smaller than the transverse axis as can be seen from Fig. 2.8, this would result in the diffraction free propagation of TM waves along the optical axis [13].

Let us now consider that the permittivity along the x-axis is infinite, the dispersion relation form (2.2) can be written as:

$$\frac{k_x^2}{\epsilon_\perp} = k_0^2 \tag{2.3}$$

This relation describes that the iso-frequency contour of the TM waves is a pair of lines with the wave vector along the optical axis. The EM energy transfer takes place perpendicular to the iso-frequency lines perpendicular to optical axis with no phase accumulation along the axis. The EM waves travelling along the optical axis would propagate with equal phase due to the flat profile of the iso-frequency contours resulting in no diffraction of waves along the axis perpendicular to the optical axis.

## 2.3.2 Applications

The property of transmission of the EM waves using infinitely anisotropic media find many applications in guiding the EM energy such as bending and splitting [13]. The diffraction free transmission of EM waves along right-angled bend formed using infinite anisotropic media takes place as shown in Fig. 2.9a. The figure shows the propagation of the TM polarized electromagnetic waves through 90° bend, the EM waves propagate along the infinite permittivity direction irrespective of the orientation. The reflection less transfer of energy through the bend happens when the two

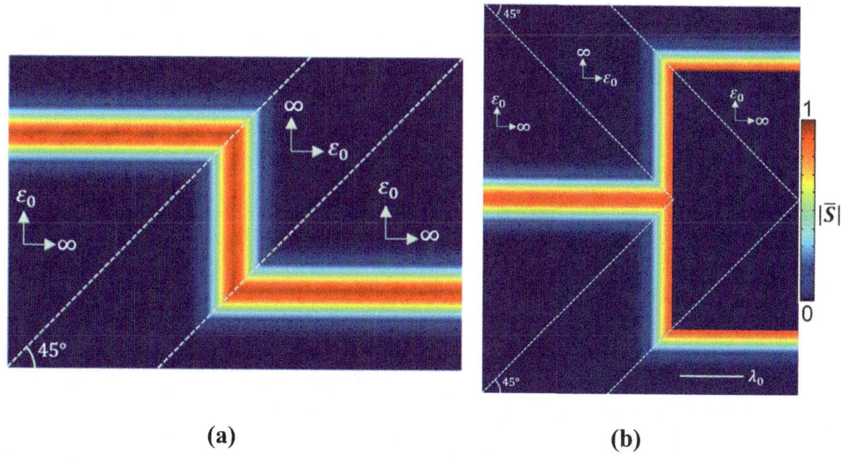


**Fig. 2.9** Manipulation of EM Gaussian beam using infinite anisotropic media, **a** bending and **b** splitting ([13], Reprinted with permission from IOP)

media have same finite permittivity component. Using this technique, reflection less EM wave splitter can also be designed using the infinite anisotropic media as shown in Fig. 2.9b. The TM polarized Gaussian beam is split into two beams right along the central line by properly choosing the anisotropic media with equal transverse permittivity values (finite).

## 2.4 Anisotropic Metamaterial Realization

### *2.4.1 Uniaxial Metamaterial Realization*

The materials with finite and transverse permittivity and extremely large longitudinal permittivity are very difficult to find in nature. However, these materials with anisotropy can be designed using metamaterials by multiple layers of high and low relative permittivity media. Considering each layers is much smaller than the operating wavelength, the effective permittivity of a multilayer structure (stacking along y-direction) with permittivity values of $\epsilon_1$ and $\epsilon_2$ occupying a fractional volume of $f$ and $1 - f$ can be given by Maxwell-Garnet theory as [14]:

$$\epsilon_x = f\epsilon_1 + (1 - f)\epsilon_2 \tag{2.4}$$

$$\epsilon_y = \frac{\epsilon_1\epsilon_2}{(1 - f)\epsilon_1 + f\epsilon_2} \tag{2.5}$$

In order to realize the extreme anisotropy with infinite permittivity along x-direction and small (finite) value of permittivity along the y-direction, we can stack metal layers having infinite permittivity with air gaps to have anisotropic media. For multilayer structure with $f = \frac{t_m}{t_{air}} \ll 1$ ($t_m$ = metal thickness, $t_{air}$ = air thickness), $\epsilon_1 = \infty$ and $\epsilon_2 = \epsilon_0$, the value of permittivity along x- and y- directions calculated from (2.4) and (2.5) are $\epsilon_x = \infty$ and $\epsilon_y = \epsilon_0$.

The value of effective permittivity for a multilayer structure with high (but finite) and low permittivity can be studied using the full wave EM simulations. For example, 8-dielectric slabs each with thickness of 0.254 mm and $\epsilon_r = 14$ and arraying these slabs with 0.254 mm air gaps is used to mimic the anisotropic media. The array of these dielectric slabs along with air gaps would result in different permittivity values along x- and transverse direction (y-direction) and can be predicted using (2.4) and (2.5). With the fractional volume occupied by the eight dielectric slabs, $f = 8/15$ we get, $\epsilon_x = 7.9$ and $\epsilon_y = 1.98$. These results can be verified also in the simulation by using perfect electric conductor (PEC) and perfect magnetic conductor (PMC) along x-direction (y-direction for $\epsilon_y$) and z-direction as shown in Fig. 2.10a. The effective permittivity can be plotted as a function of frequency for the 8-dielectic slabs with 7 air gaps (0.254 mm each) in Fig. 2.10b. The simulations results support the calculated values above. Thus, by arraying 8-diectric slabs with air gaps would

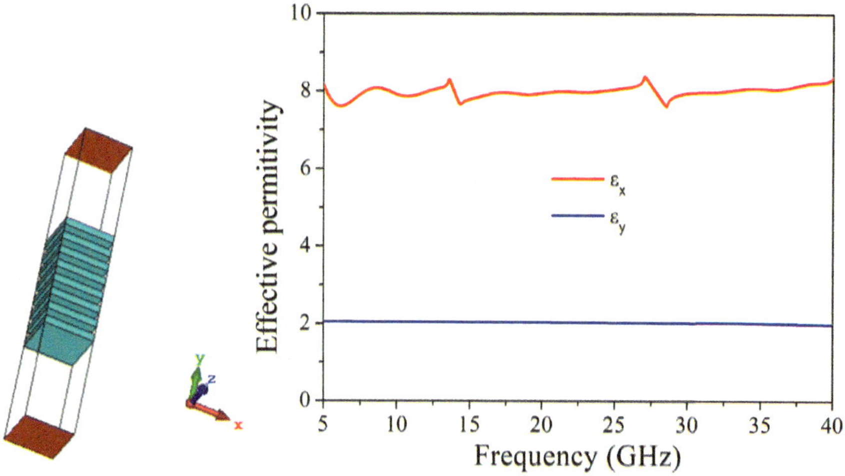


(a) (b)

**Fig. 2.10** **a** Simulation setup for effective permittivity of the 8-dielectric slabs, PEC along x-direction for $\epsilon_x$ (for $\epsilon_y$, PEC along y-direction) and PMC along z-direction, **b** extracted permittivity along x- and y-direction

mimic the bulk media with $\epsilon_x = 7.98$ and $\epsilon_y = 1.98$. However, the permittivity along the z-direction would be same as along the x-direction due to symmetry which affects the reflections.

The realization of the epsilon-high anisotropic media can be done using the array of modified asymmetric electric-LC (ELC) and the response is characterized by effective permittivity, which is controlled by the shape and size of the resonator. These ELC resonators can be configured for customizable permittivity values. Strong electric resonance results owing to the coupling of the electric field, and the desired value of the permittivity can obtained [15]. To mimic the anisotropic medium, arrays of modified ELC unit cells is designed on the Nelco NY9220 substrate with thickness of 0.254 mm as shown in Fig. 2.11a. The desired permittivity value along x-axis can be obtained by choosing the dimension of the gaps along x-direction such that the strong electric resonance is achieved at the higher frequency band. Figure 2.11b shows the extracted real part of the permittivity for two specific orientation, $\epsilon_1$ and $\epsilon_2$ using method described in [16], here we see the high permittivity value of about 15.8 is achieved when the electric field is along x-direction and a value of 2.29 is achieved for the orientation when the electric field is normal to the plane of the unit cell (along y-axis) due to no coupling of electric field. Also, the three capacitive gaps (along z-axis) in the ELC unit cell are utilized to achieve strong resonance around 28-GHz band so that transverse electric waves are stopped to enter the single-EH media to avoid mixing of modes in the anisotropic medium which becomes very difficult to analyse. Therefore, the ELC unit cell to realize anisotropic single-EH

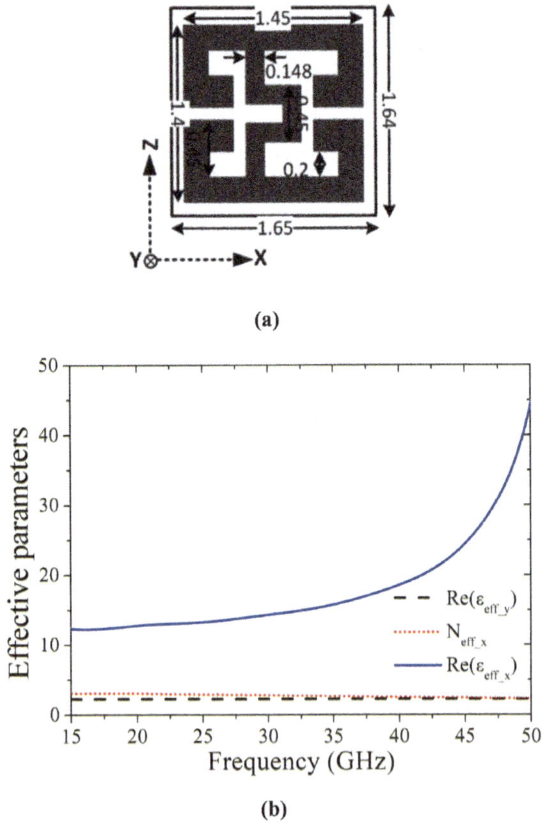


**Fig. 2.11** **a** Modified ELC-unit cell and **b** extracted permittivity of the proposed ELC unit cell for electric polarization along x- (Re ($\epsilon_x$)) and y-directions (Re ($\epsilon_y$)) along with effective refractive index ($n_x$) in x-direction (All dimensions are in millimetres)

medium for the routing of the TM waves is placed vertically with the three gaps along z-axis.

For the TM polarization, the relative permittivity along x- and y-directions can be obtained by the periodic array of modified ELC unit cell slabs with period smaller than the operating wavelength. Here we are realizing the single-EH anisotropic medium by the array of seven metamaterial ELC unit cell layers with the air gap of 0.254 mm with each layer having twelve ELC unit cells. This arrangement of ELC unit cell slabs with air gaps can be considered effective homogenous medium [10, 14] with permittivity along optic axis and transverse direction given by (2.4) and (2.5). With, $f = 7/13$, $\epsilon_1 =$ (15.8 for $\epsilon_x$ and 2.29 for $\epsilon_\perp$) and $\epsilon_2 = 1$, we get $\epsilon_x = 8.97$ and $\epsilon_\perp = 1.43$.

The high epsilon anisotropic media can also be realized using ꙅ-shaped metamaterial unit cells by stacking the high epsilon unit cell with air gaps. The designed ꙅ-shaped MTM unit cell has high epsilon value of about 15 along the x-direction and

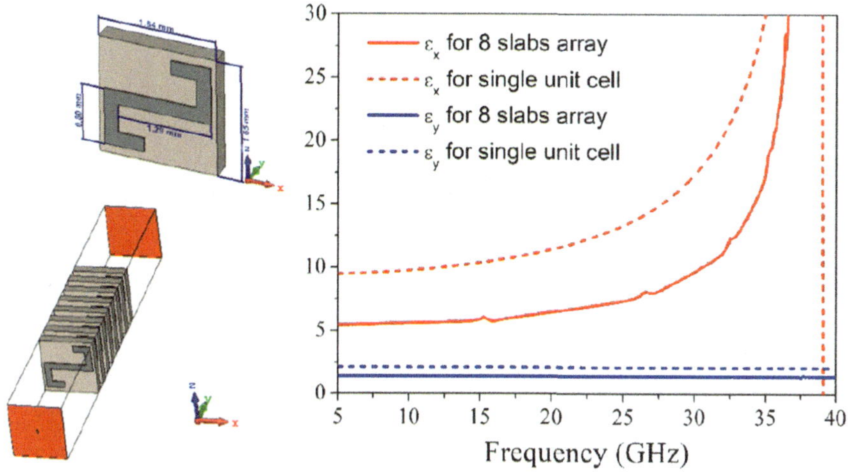


**Fig. 2.12** Ƨ-shaped MTM unit cell, 8-slab array and the extracted effective permittivity of the single unit cell and 8-slab array

2.2 along the y-direction on the Rogers RT5880 substrate. Arraying this Ƨ-shaped unit cell with seven 10 mil air gaps offers effective permittivity of 8.46 and 1.42 along x- and y-directions, respectively using (2.4) and (2.5). The Ƨ-shaped MTM unit cell and the array configuration is shown in Fig. 2.12 along with the extracted effective permittivity along the x- and y-directions of the single MTM unit cell and eight slab arrays. The simulation results verify the calculated values as shown in Fig. 2.12, individual unit cell has a permittivity of 15 along x-axis and by stacking eight slabs with low permittivity air gaps, the effective permittivity is about 9. Therefore, by arraying these MTM slabs, high epsilon anisotropic media can be realized.

### *2.4.2 Biaxial Metamaterial Realization*

The biaxial media with high and low epsilon permittivity for the orthogonal directions can be realized using anisotropic metamaterial unit cell. The anisotropic electric-LC (ELC) unit cell designed has resonant and non-resonant behavior for the two orientations which can give desired values of the permittivity for the two directions [15]. The low value of permittivity is achieved by choosing the resonance close to the frequency of interest. The resonance of the unit cell is tuned using the two capacitive gaps which results from the strong electric field coupling when the external electric field is perpendicular to the gaps. The high value of permittivity along the other direction is achieved by choosing non-resonant behavior of the ELC unit cell for the orientation with E-field parallel to capacitive plates which results in the resonance at higher frequency due to no electric field coupling to capacitive gaps. Thus, a high value of epsilon is achieved at lower frequencies in the non-resonant region.

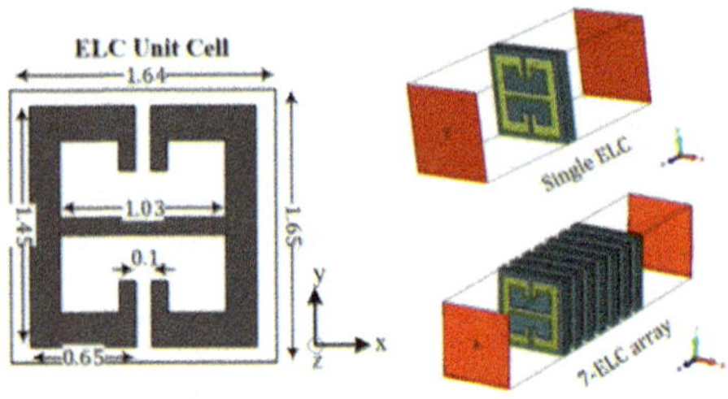


**(a)**

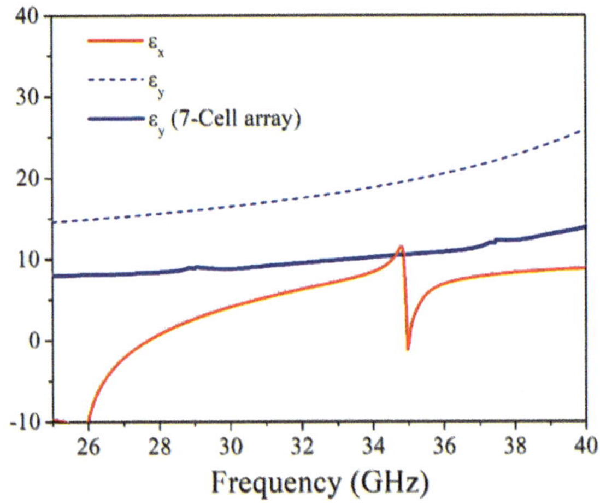


**(b)**

**Fig. 2.13** **a** Anisotropic ELC metamaterial unit cell simulation setup and **b** extracted permittivity of the unit cell for electric polarization along x- (Re($\epsilon_x$)) and y-directions (Re($\epsilon_y$)) and 7-cell array

The ELC unit cell designed at 28 GHz for high and low epsilon values is shown in Fig. 2.13a. The effective permittivity for the two directions is plotted in Fig. 2.13b, here $\epsilon_x$ represents the permittivity along the x-direction which is extracted from the S-parameters using the method given in [16] for the boundary condition PEC and PMC along x- and y-axis respectively. Similarly, $\epsilon_y$ is plotted in the same figure for the PEC and PMC conditions along y-axis and x-axis respectively. Here we can see a high value of $\epsilon_y = 15.7$ is achieved for the designed unit cell along the y-axis.

## 2.5 Conclusion

This Chapter introduces the concept of metamaterials with emphasis on the applications of these engineered materials for antenna applications. Different applications of NIM, ZIM and anisotropic metamaterials are discussed. In the later part of the Chapter, uniaxial and biaxial media along with the realization using metamaterials for practical implementation are discussed in detail.

## References


1. Wieglhofer WS, Lakhtakia A (eds) Introduction to complex mediums for optics and electromagnetics. In: SPIE, 2003. http://doi.org/10.1117/3.504610
2. Veselago VG (1968) The electrodynamics of substances with simultaneously negative values of ε and μ. Sov Phys Uspekhi 10:509. https://doi.org/10.1070/PU1968v010n04ABEH003699
3. Shelby RA, Smith DR, Schultz S (2001) Experimental verification of a negative index of refraction. Science 292(5514):77–79. https://doi.org/10.1126/science.1058847
4. Li J, Zeng Q, Liu R, Denidni TA (2017) Beam-tilting antenna with negative refractive index metamaterial loading. IEEE Antennas Wirel Propag Lett 16:2030–2033. https://doi.org/10.1109/LAWP.2017.2694400
5. Enoch S, Tayeb G, Sabouroux P, Guérin N, Vincent P (2002) A metamaterial for directive emission. Phys Rev Lett 89(21):213902. http://doi.org/10.1103/PhysRevLett.89.213902
6. Li D, Szabo Z, Qing X, Li E-P, Chen ZN (2012) A high gain antenna with an optimized metamaterial inspired superstrate. IEEE Trans Antennas Propag 60(12):6018–6023. https://doi.org/10.1109/TAP.2012.2213231
7. Rajanna PKT, Rudramuni K, Kandasamy K (2019) A high-gain circularly polarized antenna using zero-index metamaterial. IEEE Antennas Wirel Propag Lett 18(6):1129–1133. https://doi.org/10.1109/LAWP.2019.2910805
8. Christensen J, García de Abajo FJ (2010) Slow plasmonic slab waveguide as a superlens for visible light. Phys Rev B 82(16):161103. http://doi.org/10.1103/PhysRevB.82.161103
9. Schwartz BT, Piestun R (2004) Waveguiding in air by total external reflection from ultralow index metamaterials. Appl Phys Lett 85(1):1–3. https://doi.org/10.1063/1.1764596
10. Catrysse PB, Fan S (2011) Transverse electromagnetic modes in aperture waveguides containing a metamaterial with extreme anisotropy. Phys Rev Lett 106(22):223902. http://doi.org/10.1103/PhysRevLett.106.223902
11. Clemmow PC (1963) The resolution of a dipole field into transverse electric and transverse magnetic waves. Proc Inst Electr Eng 110(1):107–111. https://doi.org/10.1049/piee.1963.0016
12. Orfanidis SJ (2016) Electromagnetic waves and antennas. https://www.ece.rutgers.edu/~orfanidi/ewa/. Accessed Feb 14, 2019
13. Zhang Y, Zhang B (2017) Bending, splitting, compressing and expanding of electromagnetic waves in infinitely anisotropic media. J Opt 20(1):014001. http://doi.org/10.1088/2040-8986/aa99ad
14. Yang X, Yao J, Rho J, Yin X, Zhang X (2012) Experimental realization of three-dimensional indefinite cavities at the nanoscale with anomalous scaling laws. Nat Photonics 6(7):450–454. https://doi.org/10.1038/nphoton.2012.124
15. Schurig D, Mock JJ, Smith DR (2006) Electric-field-coupled resonators for negative permittivity metamaterials. Appl Phys Lett 88(4):041109. http://doi.org/10.1063/1.2166681
16. Szabo Z, Park G-H, Hedge R, Li E-P (2010) A unique extraction of metamaterial parameters based on Kramers-Kronig relationship. IEEE Trans Microw Theory Tech 58(10):2646–2653. https://doi.org/10.1109/TMTT.2010.2065310

# Chapter 3
# Millimeter Wave High Gain Antennas

## 3.1 Introduction

In order to meet the present day demands of high data rates with low latency, the wireless communication networks are now utilizing millimetre waves spectrum such as 28-GHz band for next generation 5G communication. However, the propagation losses that include free space path loss and atmospheric loss would result in lower signal-to-noise ratio at the receiver side, thereby reducing the range of the communication. The shorter wavelengths of millimeter waves interact with the atmospheric constituents such as oxygen and water vapours which results in losses in the signal power. However, the advantages of shorter wavelength mmWaves include packing large number of antenna elements into small form factor and larger bandwidths available at these frequency bands.

The first step in realizing the mmWave wireless communication system is to design efficient antennas which are highly directional and compact. The compact high gain antennas with suitable impedance bandwidth would be quite useful for transmission of mmWave signals over longer distances. The power received can be increased significantly using highly directional antennas at the transmitter and receiver side which would compensate for the free space path loss resulting in higher signal to noise ratio for LOS data communication.

## 3.2 Aperture Size Versus Gain

The gain of the antenna describes the amount of power transmitted/received in a particular direction (radiation peak direction) with respect to the power transmitted/received due to an isotropic source. For example, an antenna with a gain of 10 dB means the received power (in peak direction) using this antenna would be ten times more (10 dB higher) than what is received using a lossless isotropic antenna



S. K. Koul and Z. Wani, *Novel Millimetre Wave Antennas for MIMO and 5G Applications*, Lecture Notes in Electrical Engineering 819,
https://doi.org/10.1007/978-981-16-7278-1_3

(receives equally in all directions). The power received using a pair of antennas with gain $G_r$ and $G_t$ separated by distance $d$ is directly proportional to the gain of the antennas and inversely to the square of frequency of the signal. At the mmWave frequencies the received power would be small, however, the power level can be increased by using high gain antennas at both ends. The gain of the antenna depends on the effective aperture ($A_e$) of the antenna as:

$$G = k\frac{4\pi}{\lambda^2}A_e \tag{3.1}$$

where $\lambda$ is the operating wavelength and $k$ ($0 \le k \le 1$) is the antenna efficiency factor. The effective aperture describes how much power is captured and delivered to the antenna terminals by this aperture/area from a plane wave incident upon it [1]. Suppose a uniform plane wave with power density of $S(\mathrm{W/m^2})$ is incident on an antenna with aperture area of $A_e$, the power delivered to the terminals of the antenna would be given by:

$$P_t = SA_e \tag{3.2}$$

Let us take an example of a horn antenna with physical area of $A_p(= w \times h)$ at the end of the horn as shown in Fig. 3.1. We know that the electric field at the edges of the horn opening is zero, which results in non-uniform field response across the physical area. Therefore, the effective aperture area would be smaller than the actual physical area of the horn. The ratio of the effective aperture to the physical area of the antenna is known as aperture efficiency ($e_{ap} = A_e/A_p$) of the antenna.

From the relation between the aperture area and the gain in (3.1), the gain of the antenna would increase with the increasing aperture size or increasing frequency keeping the aperture area fixed. Therefore, in order to have a higher gain, the designer has to somehow increase the aperture area of the antenna. The directivity and radiation pattern of different rectangular apertures with uniform field distribution is investigated. The directivity patterns of the rectangular apertures with uniform field

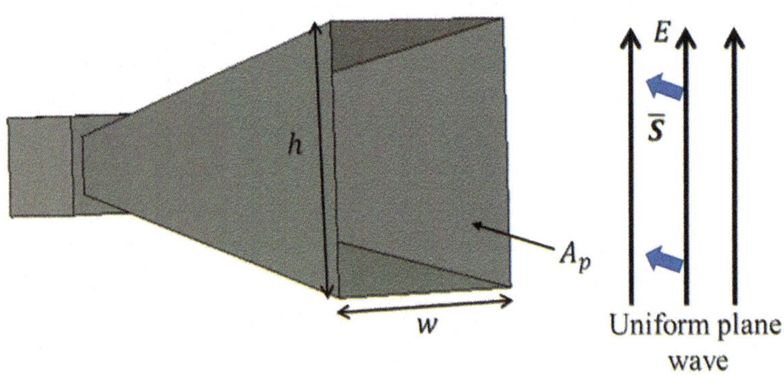


**Fig. 3.1** Uniform plane wave incident on a horn antenna

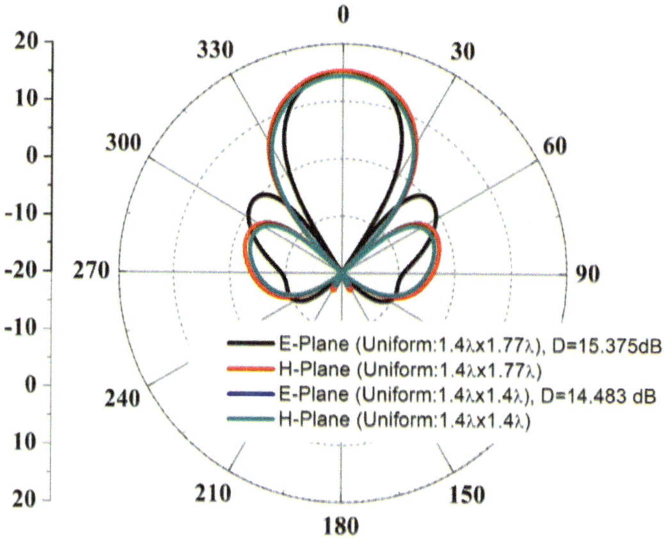


**Fig. 3.2** Radiation pattern of the rectangular aperture with uniform field distribution

distribution is shown in Fig. 3.2. With an increase in the aperture size from $1.96\lambda^2$ to $2.578\lambda^2$, the directivity increases to 15.3 dB from 14.4 dB. It is worth noting that the aperture with uniform field distribution has aperture efficiency of 1 (100%). Generally, most of the practical antennas have apertures efficiencies from 0.5 to 0.8, however there are different techniques to increase the aperture efficiency by having uniform field distribution over a physical area resulting in aperture efficiency close to 1.

## 3.3 Gain Enhancement Techniques

The gain of the antenna can be enhanced by increasing the aperture area of the antenna or by arraying the antenna elements together. The antenna array technique is the classic way of increasing the gain by increasing the effective aperture area of the antenna by utilizing number of antennas. However, the implementation of the antenna arrays at mmWaves suffer with losses and the close proximity of feed lines with radiation occurring at feed bends resulting in degraded overall performance of the antenna.

A $4 \times 4$ patch array with corporate feed on substrate having size 26.4 mm $\times$ 40 mm results in total gain of 15.5 dBi and higher side-lobe-levels in the E-plane as shown in Fig. 3.3. However, increasing the aperture size of the antenna would not be feasible for the mmWave 5G communication systems.

**Fig. 3.3** **a** E and H-plane radiation pattern of the 4 × 4 patch array with corporate feed, **b** 3D radiation pattern

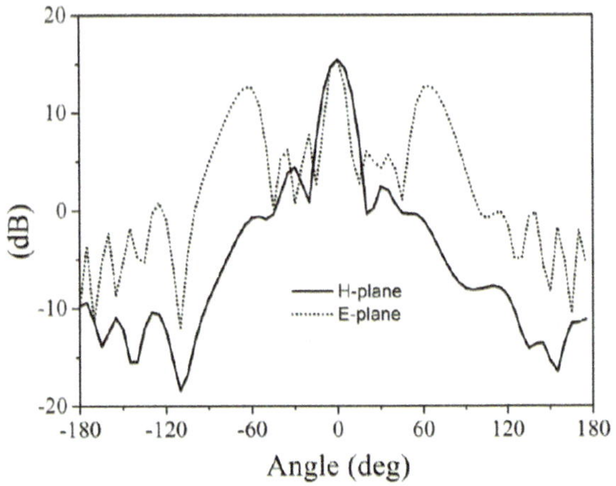


**(a)**

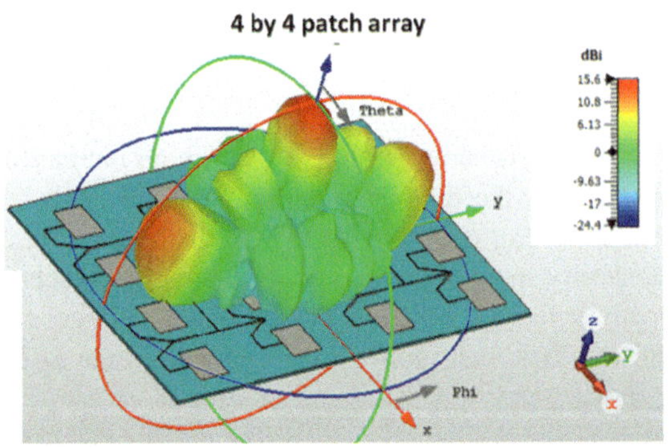


**(b)**

Several new gain enhancement approaches have been explored in recent years; resonant cavity antennas (RCA) which include partially reflecting surfaces, Fabry-parrot cavity antennas [2] and bandgap resonator antennas [3] have gained much attention due to their easy implementation with the planar technology and they offer good alternative for array antennas. The gain enhancement is achieved using partially reflective surfaces (PRS) in [4–6], however these antennas have limited impedance bandwidth which depends upon the gap between the antenna and PRS. The partially reflecting meta-surfaces which are placed at a distance of $h = \frac{\lambda_0}{2}$ ($\lambda_0$ is free space wavelength at the operating frequency) from the radiating antenna resulting in increased size of the radiating system [2, 5, 7]. Also, these techniques have low aperture efficiency and limited bandwidth [4]. In order to achieve gain enhancement over a broad frequency range, a flat dielectric slab with holes to compensate for the propagation phase is proposed in [8], however the gain enhancement is limited to merely 2.3 dB. Recently, gain enhancement is done using epsilon/mu-near-zero medias (ENZ/MNZ) in [9–11], however design of these media does affect the input impedance of the antenna resulting in narrow bandwidth. Existing methods to enhance the gain of an antenna also include employing epsilon-near-zero (ENZ)/mu-near-zero (MNZ) materials which enhance the gain in one of the orthogonal planes viz. H-plane or E-plane [12]. In [12], a zero-index anisotropic lens is designed for gain enhancement in both E- and H-planes, however the radiation bandwidth is small and is limited due to the nature of zero-index metamaterials. Also, refractive index of the zero-index media is very sensitive to small changes in the permittivity around zero. It is also required that these materials are impedance matched to the antenna [12, 13].

The near-field phase correction technique has been developed to enhance the gain of the aperture antennas. This technique is an alternative to traditional lenses in which the transmission phase is modulated using array of small phase delay unit cells [14]. The technique of near-field phase correction reported in [15–17] significantly enhances the antenna gain. The phase correcting structures proposed in [15, 16] uses array of phase shifting unit cells to locally modify the phase of the near-field radiation. However, the phase errors due to the large size of these phase shifters designed at single frequency limits the aperture efficiency to about 30% only. Also, these techniques result in large antenna size with narrow frequency of operation. A wide-band phase correcting technique reported in [17] has about 46% impedance bandwidth and 70% aperture efficiency. A 3D printed phase correcting superstrate for resonant cavity antenna is proposed in [18]. Authors have used a 3D printed dielectric structure with varying height profile to provide different transmission delays for the phase correction, however the antenna has a broadside height of $1.46\lambda_0$ which results in bulky antenna system and the antenna with superstrate has 46.4% aperture efficiency only.

## 3.4 Stacked Dielectric Loading for Gain Enhancement

The previous section describes various gain enhancement techniques such as PRS, RCA and ENZ/MNZ media, however all these techniques are resonant and the behavior is quite sensitive to the operating frequency. This section presents a non-resonant technique to enhance the gain of a microstrip patch antenna (MPA) using stacked dielectric slabs loaded on the top [19]. The design of the superstrate uses the concept of phase compensation to achieve equal phase profile by decreasing the effective permittivity along the boundary of the superstrate. The fabricated all-dielectric stacked superstrate is loaded to the MPA and the measured results are presented in the following sections.

### 3.4.1 Design and Working Mechanism

The configuration of the stacked 8-dielectric slabs with air gaps along the boundary and no air gaps in the center is shown Fig. 3.4. The dielectric slabs are 0.254 mm thick and have relative permittivity of 2.2. The effective permittivity of the dielectric slabs can be controlled by adjusting the air gaps between them. The stacked 8-dielectric slabs ($\epsilon_{rd} = 2.2$) with 0.254 mm air gaps in-between has effective permittivity less than 2.2, this arrangement of slabs with air gaps can be regarded as homogenized media with effective permittivity given by (2.4). For, $f = \frac{8}{15}$ the fraction of the volume occupied by the 8-dielectric slabs [20], the effective permittivity for the 8-layer stacked slabs with $\epsilon_{rd} = 2.2$ and seven air gaps in between results in, $\epsilon_{eff} = 1.6$. The effective permittivity of the stacked slabs with air gaps is studied in simulations by employing PEC and PMC boundary condition to excite plane wave for extraction of the media parameters as shown in Fig. 3.5a. The effective permittivity is plotted in Fig. 3.5b for the 8-dielectric slabs with air gaps, the decrease in effective permittivity value is observed from 2.2 to 1.6. Thus, by choosing appropriate air gaps the effective permittivity of the medium decreases. The phase compensation for the electromagnetic waves which travel along the oblique angle from the broadside direction would have more path length than along the broadside direction as shown in Fig. 3.4. Therefore, by choosing a low permittivity of the loading media along the boundary of the superstrate would result in less propagation phase. The phase difference ($\delta\xi$) between the EM rays in the center of the stacking (high permittivity media, $n_1$) and in the boundary of the media (lower permittivity media, $n_2$) is given as:

$$\delta\xi = \frac{2\pi}{\lambda_0} d(n_1 - n_2) \tag{3.3}$$

where $d$ is the thickness of the media and $\lambda_0$ is the free space wavelength at 28 GHz. For $n_1 = 1.48, n_2 = 1.26$; $\delta\xi = 28.2°$, thus by employing stacked media with the

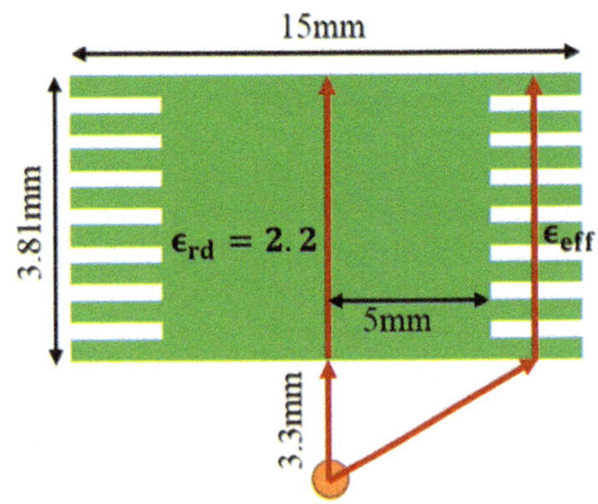


**Fig. 3.4** Proposed all dielectric stacked superstrate for phase correction using eight dielectric slabs ([19], Reprinted with permission from IEEE)

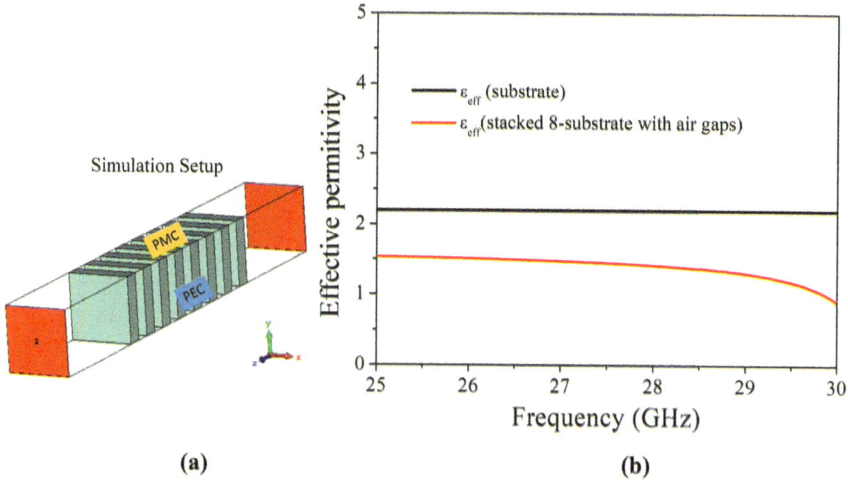


**Fig. 3.5** **a** Simulation setup and **b** the extracted effective permittivity of the 8-stacked dielectric slabs with 7-air gaps ([19], Reprinted with permission from IEEE)

boundary results in the phase compensation of 28.2° which would compensate for the larger path length of the oblique EM rays.

There are no air gaps between the dielectric slabs in the center of the superstrate in order to achieve higher effective permittivity along the broadside direction than the boundary of the loading media. The higher permittivity would introduce more propagation phase which would compensate for the larger path lengths along oblique direction from the broadside direction. The stacked 8-layer media is loaded at 3.3 mm to the MPA as shown in Fig. 3.6a. The phase of the electric field ($E_y$-component) radiated from the MPA is plotted along the x-axis at the cut plane z = 7.5 mm (top of the stacking media) with stacking and without stacking in Fig. 3.6b, we can observe that the electric field profile is flatter for the dielectric stacking case due to the phase compensation by employing lower permittivity along the boundary of the stacking media. This phase compensation of 40° would result in the gain enhancement of the antenna.

The phase distribution of the y-component of the electric field in xz-plane (y = 0 mm) and xy-plane (z = 0.754 mm) without loading and with stacked superstrate is shown in Fig. 3.7. It can be seen from the figure that the phase distribution in case of loading with stacked superstrate is more uniform in the xy-plane than without loading.

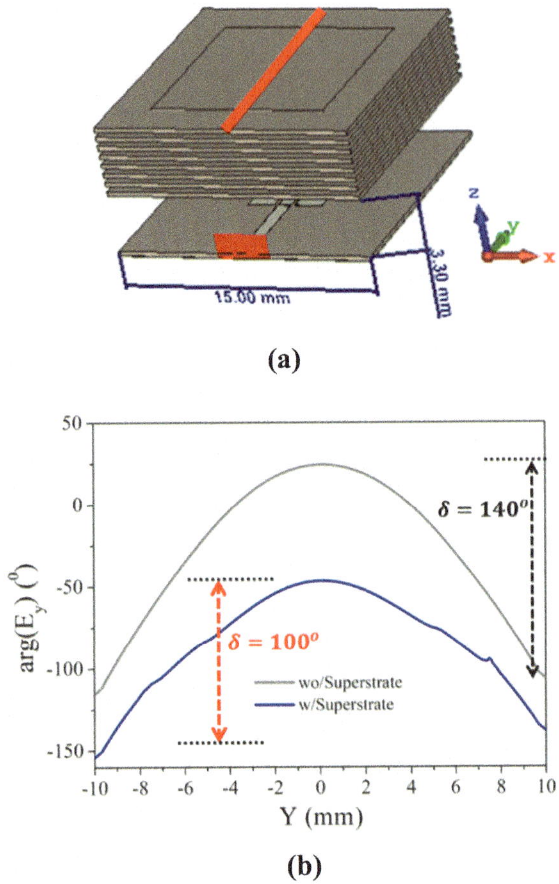


**Fig. 3.6** **a** Configuration of MPA with 8-stacked dielectrics, **b** phase of y-component of electric field along x-axis for z = 0.75 mm, exit boundary ([19], Reprinted with permission from IEEE)

### *3.4.2 Experimental Results*

The MPA is designed at 28-GHz on a Neltec substrate having relative permittivity of 2.2. The stacking media with no airgaps in the center and airgaps along the boundary is achieved by sandwiching smaller 10 mil substrate ($10 \times 10\,\text{mm}^2$) in between the two bigger dielectric slabs ($15 \times 15\,\text{mm}^2$). The stacking superstrate contains eight bigger dielectric slabs supported by seven smaller slabs in the center. The stacking media is loaded to the MPA and is supported by a 3D printed stand, the photograph of the whole assembly is shown in Fig. 3.8. The MPA is fed using a 1.85 mm end-launch connector to measure the input reflection coefficient ($|S_{11}|$). The measured and simulated reflection coefficients are plotted in Fig. 3.9 for the MPA with superstrate. It is evident from the figure that the dielectric stacking does not affect the input reflection coefficient of the antenna. The proposed antenna has 2.5% bandwidth at a resonant frequency of 28-GHz.

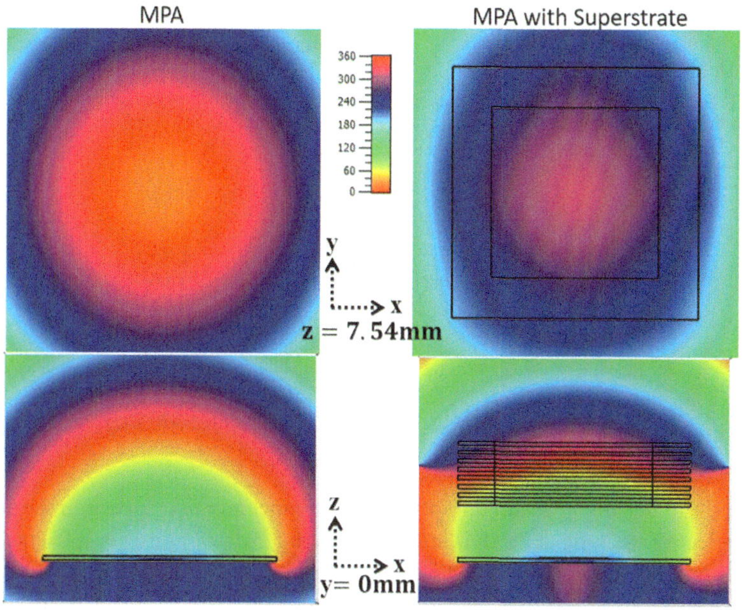


**Fig. 3.7** Phase of y-component of electric field at $\omega t = \pi/2$ in xy-plane and xz-plane for MPA only and MPA with superstrate ([19], Reprinted with permission from IEEE)

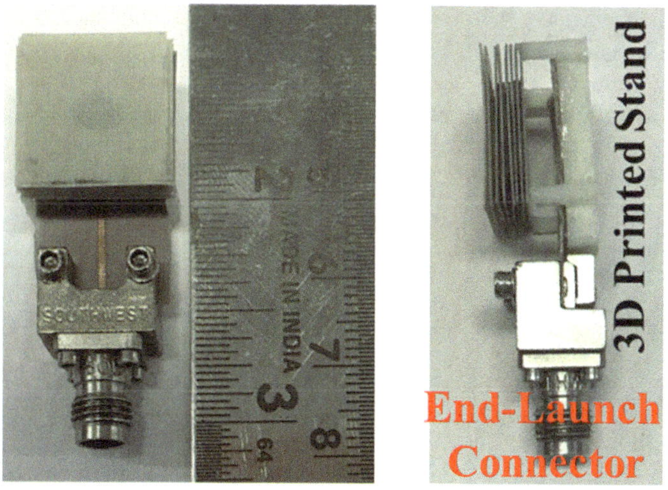


**Fig. 3.8** Top and side view of the fabricated MPA with stacked superstrate ([19], Reprinted with permission from IEEE)

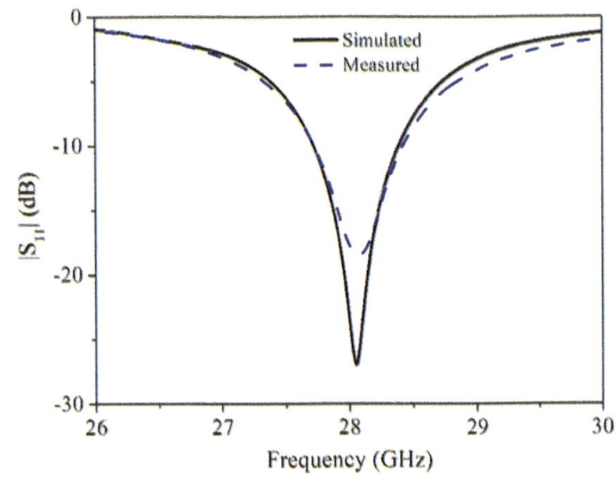


**Fig. 3.9** Simulated and measured input reflection coefficient of the proposed antenna ([19], Reprinted with permission from IEEE)

The simulated and measured H-plane and E-plane radiation pattern of the antenna with dielectric stacking at 28 GHz are plotted in Fig. 3.10. As observed, the measured results agree well with the simulated results, a front to back ratio more than 20 dB with a side lobe level less than −18 dB in both the planes is achieved. The peak gain of the MPA without stacked dielectric is 7.4 dB at 28 GHz and the peak gain is enhanced to 11.5 dB with stacking. The gain of the MPA with stacking bigger substrate only over the patch (without smaller substrates in the center) is about 8.9 dBi. Therefore, a gain enhancement of 4.1 dB is achieved using the stacking superstrate (smaller as well as bigger substrates). The simulated and measured gain plots are shown in Fig. 3.11 for the MPA with stacking superstrate. Superimposed on this graph is the gain of the MPA only. This technique of gain enhancement offers advantage in terms of simple design, non-resonant behavior of the media and no conductor losses in the stacking superstrate.

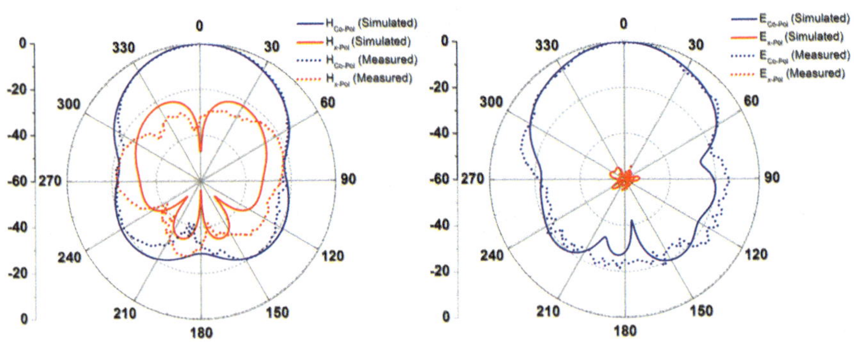


**Fig. 3.10** Measured and simulated H-plane and E-plane radiation pattern of the antenna with stacked superstrate at 28-GHz (solid: simulated, dashed: measured) ([19], Reprinted with permission from IEEE)

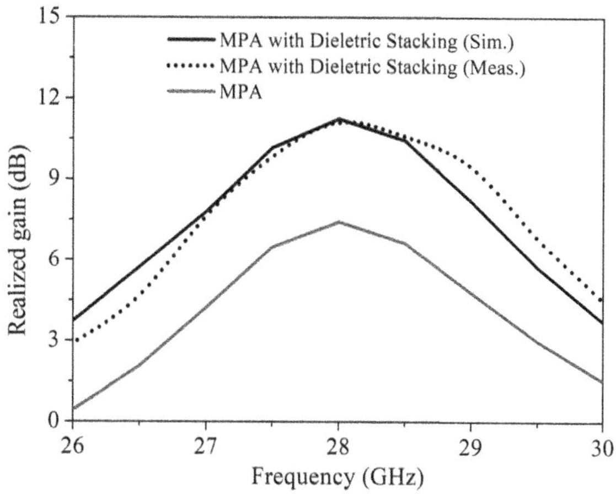


**Fig. 3.11** Measured and simulated gain comparison of the proposed antenna with stacked superstrate and MPA without stacking ([19], Reprinted with permission from IEEE)

## 3.5 Gain Enhancement Using High-Low Epsilon Biaxial Anisotropic Lens

The gain enhancement technique presented in previous section using the dielectric stacked loading results in an aperture efficiency of 57.2% only. To achieve higher aperture efficiency close to 100%, this section presents the technique for gain and aperture efficiency enhancement using the biaxial anisotropic slab with high and low epsilon values along the two orthogonal directions. First the mechanism of gain enhancement is discussed with the high/low epsilon (HLE) slab and then its practical realization using metamaterials for millimeter wave antennas is presented.

### *3.5.1 Working Mechanism and Anisotropic Lens*

Let us consider a biaxial anisotropic slab with permittivity given by tensor, $\epsilon = \epsilon_0 \left[ \epsilon_x \; \epsilon_y \; \epsilon_z \right]$ and permeability, $\mu_0$. Where $\epsilon_0$ and $\mu_0$ are the free space permittivity and permeability respectively and $\epsilon_x$, $\epsilon_y$, $\epsilon_z$ are the relative permittivity along x-, y- and z-directions respectively. Electromagnetic waves with polarization along y-axis are incident on this biaxial anisotropic media. Propagation inside the anisotropic media for TM and TE polarized waves in $yz$-plane and $xz$-plane respectively is governed by the following dispersion relations [21]:

$$\text{TM wave: } \frac{k_y^2}{\epsilon_z} + \frac{k_z^2}{\epsilon_y} = k_0^2 \tag{3.4}$$

$$\text{TE wave: } k_x^2 + k_z^2 = k_0^2 \epsilon_y \tag{3.5}$$

where $k_x$, $k_y$ and $k_z$ are the propagation vectors along x-, y-and z-axis respectively of the medium and $k_o = \omega\sqrt{\mu_o \epsilon_o}$ is the propagation constant in free space.

For the TM wave in $yz$-plane having nonzero field components $H_x$, $E_y$ and $E_z$ the propagation along y- and z-direction will depend upon the relative permittivity along the z- and y-directions respectively. If we consider, $\epsilon_y > \epsilon_z$, we can confirm from (3.4) that $k_y < k_z$, therefore the total wave vector strongly depends upon $k_z$ and negatively on $k_y$. Thus an increase in $k_y$ results in decrease in the total wave vector. For $k_y \neq 0$ (which implies oblique propagation in the media with a larger propagation path length), the total wave vector decreases due to large $k_y$. The decrease in total wave vector will compensate for the larger path lengths. This can be verified from the plot in Fig. 3.12a between normalized total k-vector ($k/k_z$) versus the angle from the z-axis. We can see from the figure that the k-vector decreases for large oblique propagation. This can compensate for the large path length and the phase accumulation while propagating along z-axis which results in a flat phase profile of the fields in $yz$-plane, hence gain enhancement in the E-plane.

For TE waves in $xz$-plane, the total k-vector is constant and is larger than in free space as evident from (3.5). The TE fields entering in the media along the z-axis would propagate in the media and the slab can support different TE modes of propagation depending upon the dimension of the media along the x-axis and permittivity along y-axis [22]. At the exit boundary between the HLE media and free space, the tangential wave vector ($k_x$) is continuous. As the magnitude of the k-vector in free space is smaller than in the media, as can be seen from the iso-frequency contour at the exit boundary of the media shown in Fig. 3.12b. The iso-frequency graphs for the TE wave is a circle with radius in media (red) which is larger than in free space (dotted blue line) due to high epsilon along the y-axis. Therefore, most of the TE waves with smaller $k_x$ in the media would propagate and exit the media along the normal direction (z-axis), which results in radiation along the z-axis.

To study the physics of the high-low-epsilon (HLE) biaxial anisotropic media having high and low permittivity values along two orthogonal axes, first the biaxial slab with user-defined permittivity tensor and dimensions ($l_x, l_y, t_m$) is studied using plane wave excitation in full wave EM simulator. The thickness of the medium is chosen to be multiples of $\lambda_g/2$ for reflection less transmission of waves at 28 GHz [24]. The variation of power distribution in yz-plane for different values of $\epsilon_y$ is plotted in Fig. 3.13. It is observed that focusing of the power along the center of the slab takes place till $\epsilon_y = 10$ and decreases thereafter. The value of power distribution decreases for permittivity, $\epsilon_y = 12$ due to reflections occurring since the guided wavelength, $\lambda_g = 3$ mm becomes smaller than the slab thickness, $t_m = 3.55$ mm. The focusing property of the HLE slab with $\epsilon_y = 10$ and $\epsilon_x = 0.1$ with dimensions, $l_x = l_y = 15$ mm and $t_m = 3.55$ mm, is verified by the power distribution in the xz- and yz-planes. The snapshot of the power flow through the slab when excited using plane wave is shown in Fig. 3.14. We can observe that more power is distributed along the center of the slab in both the planes when excited with a plane wave on the other side of the slab. Thus, the HLE slab work as converging lens which when excited with a plane wave on one side, converges the power along the center of the

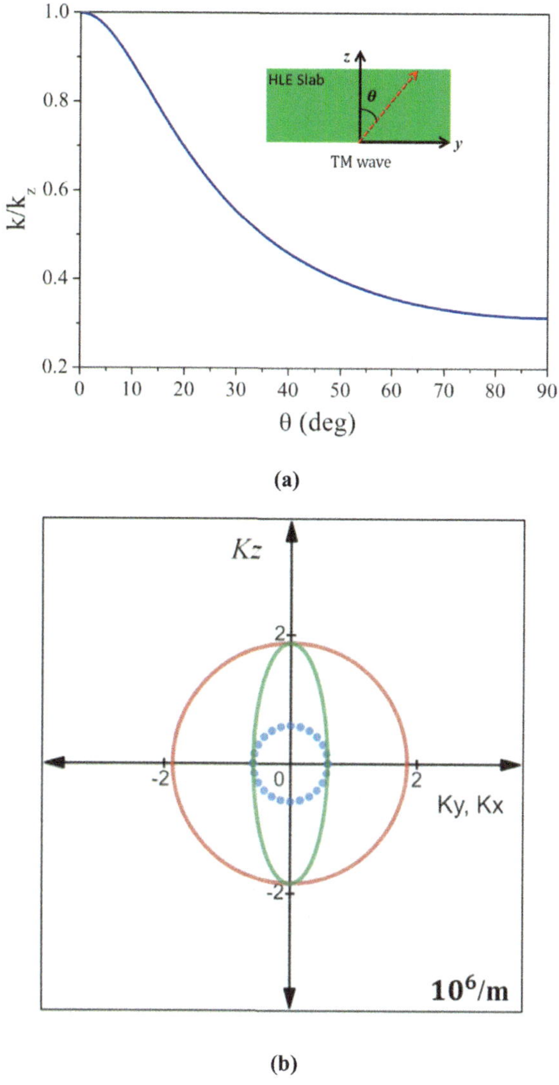


**Fig. 3.12** **a** Normalized k-vector versus angle from the z-axis for TM waves in yz-plane, **b** the iso-frequency contour for TM waves in yz-plane (green curve), TE waves in xz-plane (red curve) and for free space in xz-plane (dotted blue)

slab on the other side. This property of the HLE slab is used for gain enhancement of an antenna in the following sections.

### 3.5.2 *Gain Enhancement Using Anisotropic HLE Slab*

The propagation of fields through the HLE slab and the gain enhancement is studied using the full wave EM simulations. The simulations are performed using time

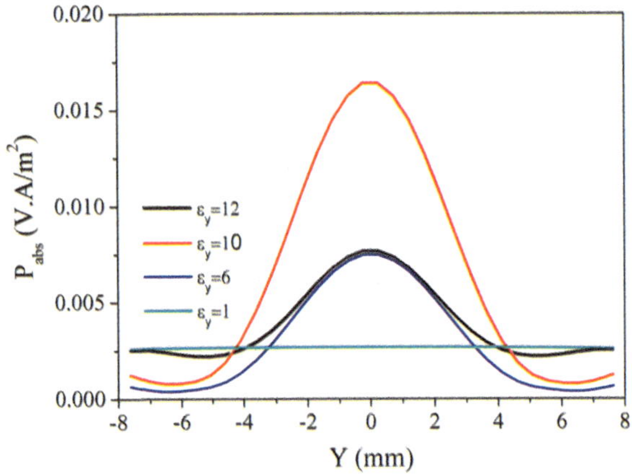


**Fig. 3.13** Power distribution through the anisotropic slab when excited with a plane wave for different permittivity values for cut plane, $z = 5.8$ mm along y-axis ([23], Reprinted with permission from IEEE)

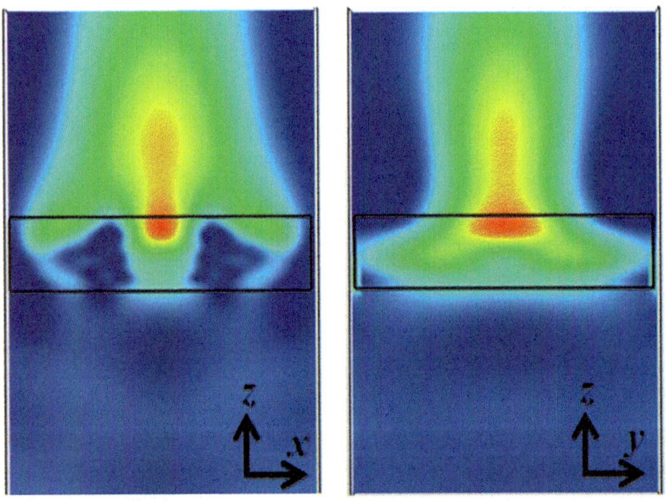


**Fig. 3.14** Power flow distribution through the HLE slab when excited with a y-polarized plane wave ([23], Reprinted with permission from IEEE)

domain solver, a horn antenna designed at Ka-band is used for studying the effect of HLE slab on the radiated fields from the horn. The schematic of the horn antenna with the HLE slab is shown in Fig. 3.15, the anisotropic slab with permittivity tensor [$\epsilon_x$, $\epsilon_y$ and $\epsilon_z$] is placed in front of the horn antenna. The E- and H-fields from the horn antenna couple to the HLE slab and the snapshot of the transverse electric and magnetic fields is shown in Fig. 3.16 for the permittivity tensor [0.1, 10, 1]. It is observed that the E- and H-fields from the horn are transformed to plane waves after

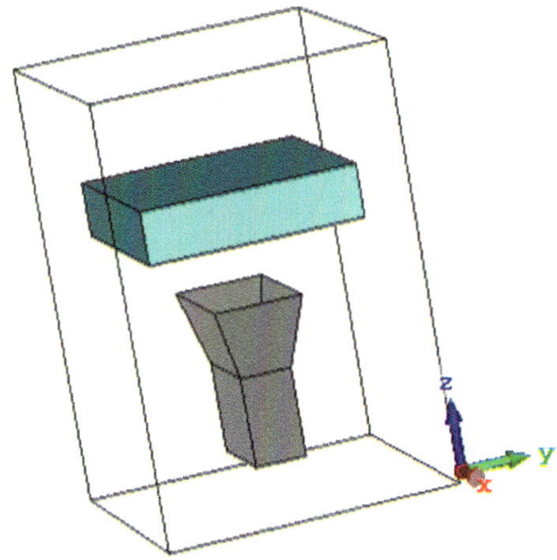


**Fig. 3.15** Horn antenna loaded with HLE slab

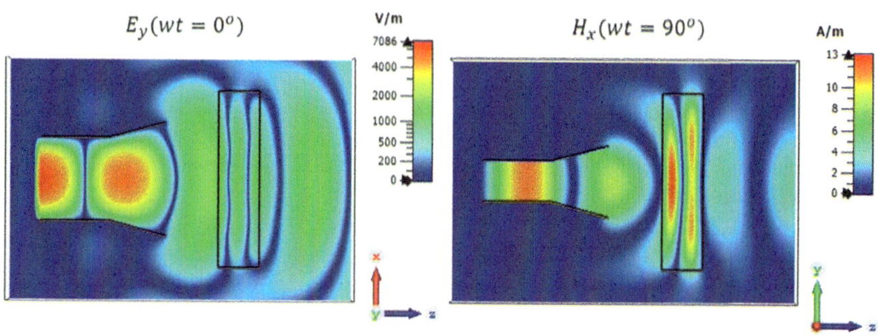


**Fig. 3.16** Snapshot of the transverse electric and magnetic fields

entering the anisotropic HLE slab resulting in the gain enhancement. The gain plot for the various permittivity values is plotted in Fig. 3.17. It is observed that one gets enhanced gain for higher epsilon value of $\epsilon_y$ and lower value of $\epsilon_x$ when Ka-band horn loaded with anisotropic HLE slab at 28 GHz.

The gain enhancement resulting from the loading of the horn antenna with the HLE slab is shown in Fig. 3.18. For comparison, the gain of the horn antenna without loading is also plotted in Fig. 3.18. We can see that the wideband gain enhancement is achieved by loading horn antenna with HLE slab. To further verify the theory of gain enhancement using HLE slab, a microstrip patch antenna (MPA) with $w = 4.4$ mm and $l = 3.44$ mm is designed on 10 mil thick substrate ($\epsilon_d = 2.2$) at 28 GHz using CST Microwave studio, the HLE slab with user-defined material is placed above the patch antenna at a distance of $h = 3.5$ mm ($= \lambda_0/3$ at 28 GHz). For the gain enhancement study, the designed MPA is loaded with the high-low-epsilon (HLE) slab having permittivity tensor [0.1, 10, and 1]. The E- and H-fields from the MPA along y- and x-direction respectively couple to the HLE slab loaded over the antenna.

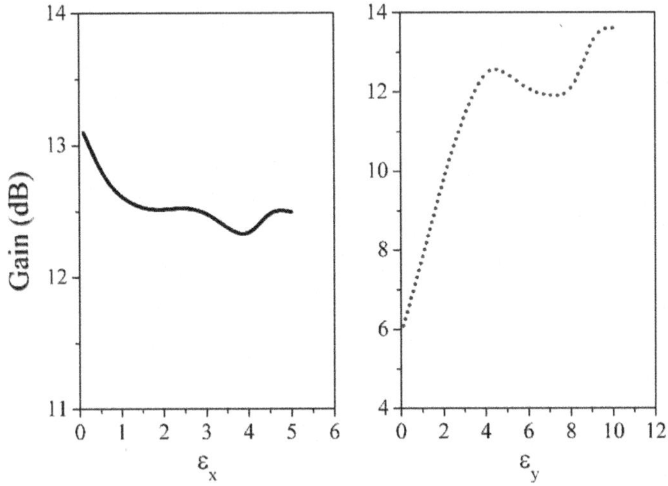


**Fig. 3.17** Variation of gain versus epsilon values along x- and y-directions

**Fig. 3.18** Gain comparison of the horn antenna with HLE slab depicting the wide band gain enhancement

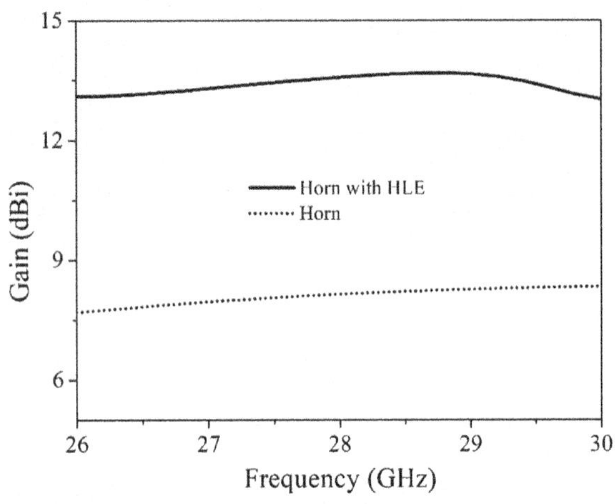


The TM fields in *yz*-plane couple to the media with no phase accumulation along the y-axis resulting in gain enhancement. The input reflection coefficient of the MPA with HLE slab loading is shown in Fig. 3.19.

It is observed from the figure that the input reflection coefficient ($|S_{11}|$) remains almost unchanged with loading of the MPA with HLE slab very close to the antenna. Thus, we can load the HLE slab very close to the antenna for guiding the electromagnetic fields along the z-axis. Figure 3.20 shows the y-component of electric field and x-component of magnetic field in xz- and yz-planes respectively. The plot shows the flat profile of the $H_x$ and $E_y$ fields for the various values of $\epsilon_y$ and $\epsilon_x$ respectively. This flat profile of the transverse field components is due to higher and lower values

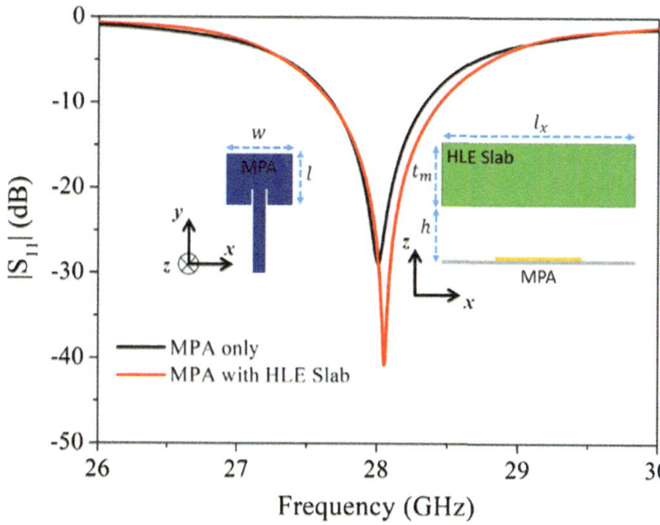


**Fig. 3.19** Configuration of the microstrip patch antenna loaded with HLE slab and input reflection coefficient of the MPA with and without HLE slab loading ([23], Reprinted with permission from IEEE)

for $\epsilon_y$ and $\epsilon_x$ respectively. For the transverse electric and magnetic field in $xz$-plane and $yz$-plane respectively, the phase profile is flat along the y- and x-direction for $\epsilon_y = 10$ and $\epsilon_x = 0.1$, with overall propagation along the z-direction, hence the gain enhancement in H- and E-plane respectively.

The gain pattern in yz-plane of the MPA with the HLE slab is plotted in Fig. 3.20 for different thickness of the slab for the permittivity tensor, [0.1, 10 and 1]. For TM waves in yz-plane, the effective refractive index varies from 3.16 to 1 for incident angles 0° to 90° (as the refractive index of TM waves in an anisotropic media depends upon the direction of propagation). For, $\theta = 0°$ means propagation along z-axis and E-field is in y-axis, refractive index is $\sqrt{\epsilon_y}$). For reflection-less propagation of the waves through the slab, the thickness should be multiples of half guide wavelength, $t_m = m\frac{\lambda_g}{2}$. Thus, we can observe from Fig. 3.21, gain increases when the length of the slab approaches $t_m = \lambda_g = 3.55$ mm. When the thickness of the media exceeds $\lambda_g$, total transmission through the slab decreases resulting in gain drop as can be seen from the gain curve for $t_m = 4.6$ mm from the figure. Figure 3.22 shows the 3D radiation pattern of MPA with the HLE slab loading resulting in a peak gain of 13.9 dBi with 26.6° and 39.6° beamwidth in E- and H-plane respectively.

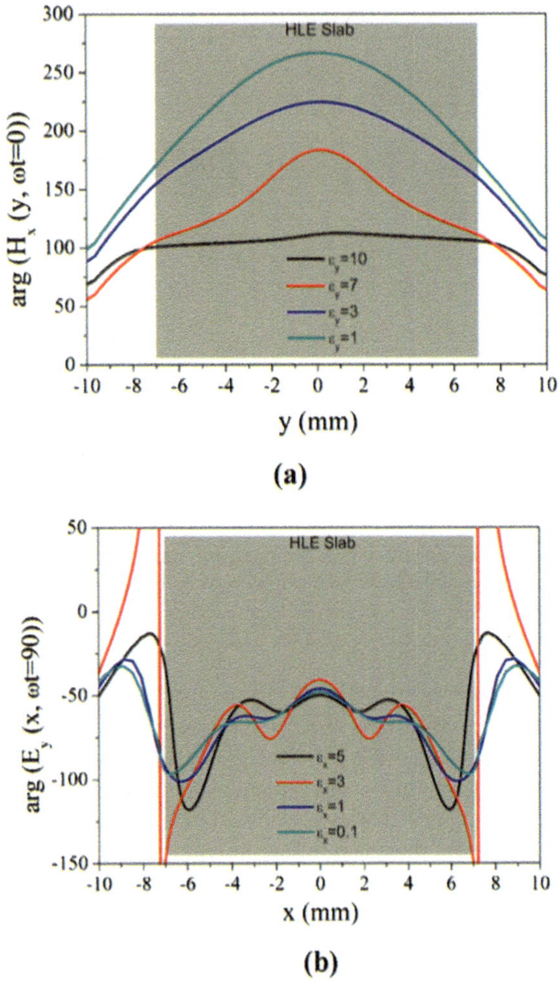


**Fig. 3.20** Phase variation of transverse E- and H-field components along y- and x-axis in HLE slab, **a** phase of $H_x$ along y-axis for different $\epsilon_y$, **b** phase of $E_y$ along x-axis for different $\epsilon_x$ ([23], Reprinted with permission from IEEE)

### 3.5.3 *Antenna Gain Study and MTM Realized HLE Media*

The HLE slab for the gain enhancement presented above is practically realized using metamaterials as discussed in Chap. 2. The HLE slab having permittivity tensor [0.1, 10] is practically realized using seven stacked metamaterial (MTM) layers with each layer consist of 9 × 9 ELC unit cells and 10 mil air gaps. The implementation of the HLE lens using metamaterials is done using the stacking of seven metamaterial layers fabricated using photolithography and the photograph of the HLE lens realized using seven stacked metamaterial layers is shown in Fig. 3.23. To have the air gaps between the layers, the MTM layers at corners are supported using a 10 mil thick small substrate pieces and glued just to hold the layers together.

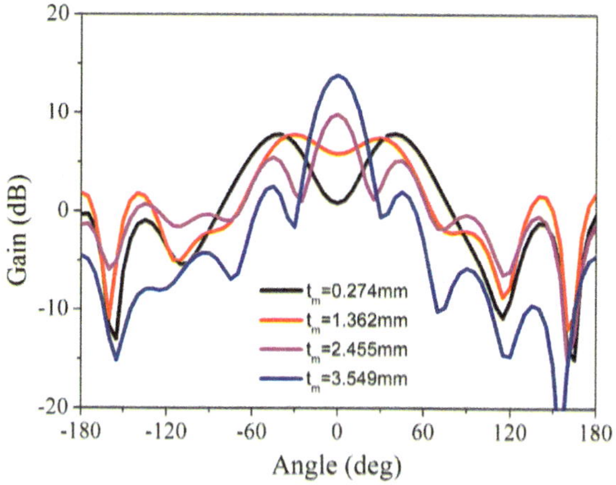


**Fig. 3.21** E-plane gain pattern of the antenna with HLE slab at 28 GHz for different slab thickness ($t_m$) ([23], Reprinted with permission from IEEE)

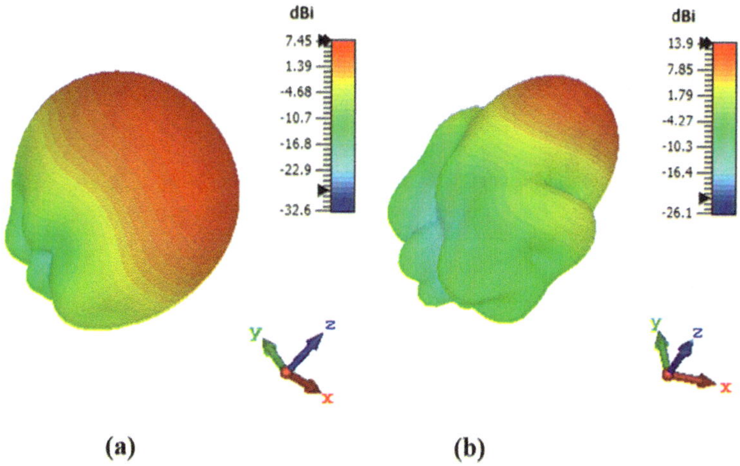


**Fig. 3.22** 3D radiation pattern of the **a** MPA and **b** MPA with HLE slab at 28 GHz

The MTM realized HLE media is loaded to the patch antenna designed at 28 GHz as shown in Fig. 3.24. Here the MTM lens consists of seven MTM layers with six air gaps which corresponds to a thickness of 3.51 mm and aperture size $14.7 \times 14.86\ mm^2$ which is the same as the HLE slab thickness used in previous section. The results obtained with MTM lens is compared with the results obtained with HLE slab. The input reflection coefficient of the patch antenna with MTM lens is shown in Fig. 3.25, the loading with the MTM lens does not affect the impedance of the patch when loaded at the same height as HLE slab. The E-plane radiation pattern of the patch antenna with different number of MTM layers loaded to the antenna is shown

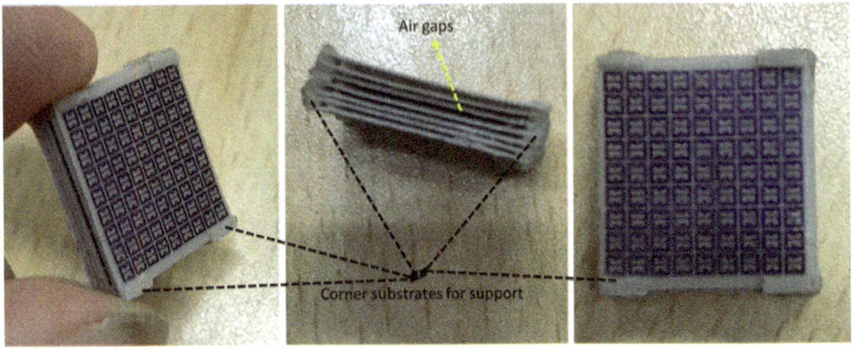


**Fig. 3.23** Metamaterial realized anisotropic HLE lens

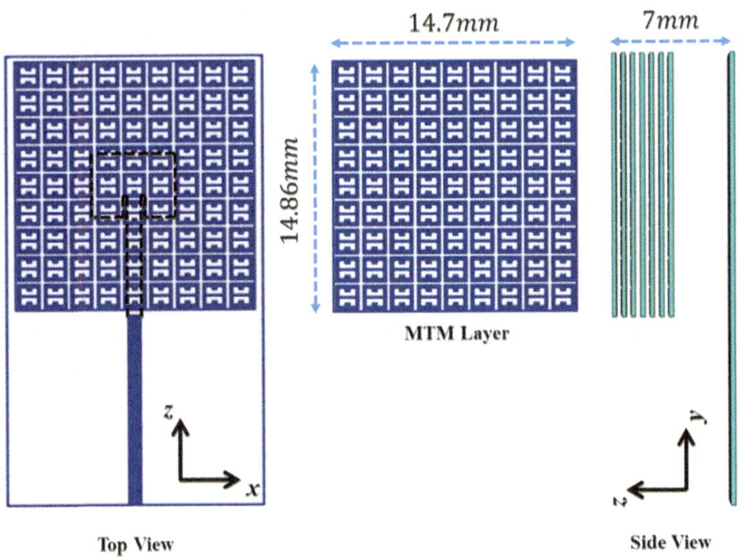


**Fig. 3.24** Schematic of the microstrip patch antenna loaded with MTM lens realized using stacked layers ([23], Reprinted with permission from IEEE)

in Fig. 3.26. We can observe the same trend in the pattern as was seen with HLE slab with different thickness in Fig. 3.21. It is also observed that the gain increases with an increase in the number of layers which is equivalently same as increasing the HLE media thickness over the patch antenna which is essential for phase correction in the media.

The poynting vector depicting focused radiation is shown in Fig. 3.27 for the patch antenna only, MPA with HLE slab and MTM lens. The figure shows the focused broadside radiation along z-direction due to the high epsilon along the y-direction.

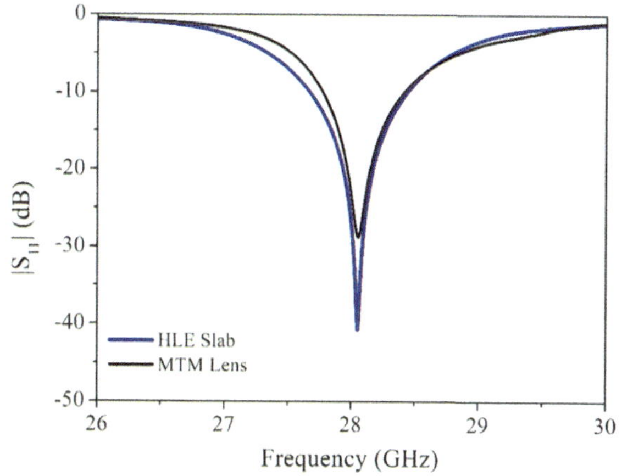


**Fig. 3.25** Input reflection coefficient ($|S_{11}|$) of the patch antenna loaded with MTM lens and with HLE slab ([23], Reprinted with permission from IEEE)

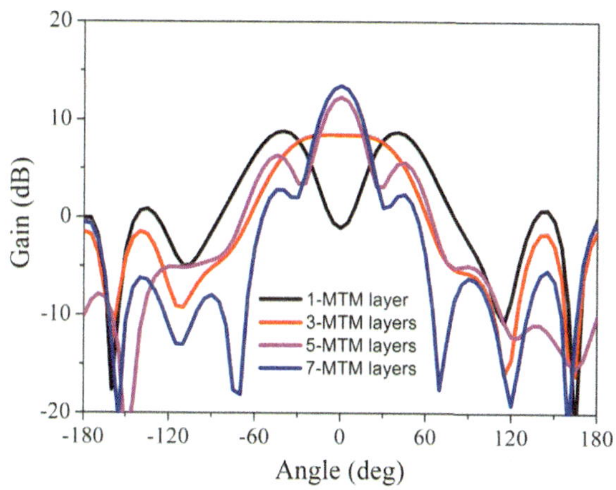


**Fig. 3.26** E-plane radiation pattern of patch antenna with varying number of MTM layers loading ([23], Reprinted with permission from IEEE)

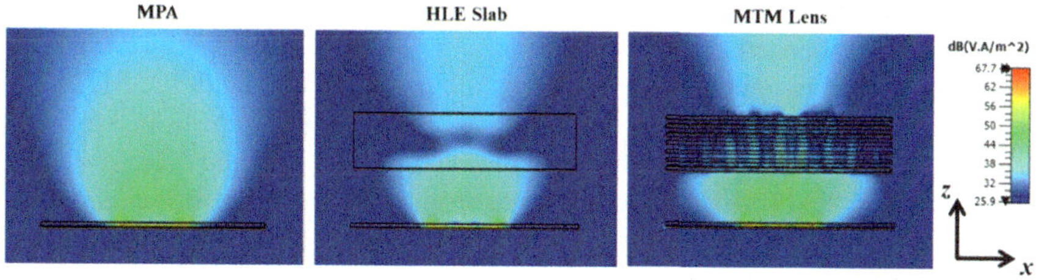


**Fig. 3.27** Power flow distribution for MPA with HLE slab and MTM lens in xz-plane ([23], Reprinted with permission from IEEE)

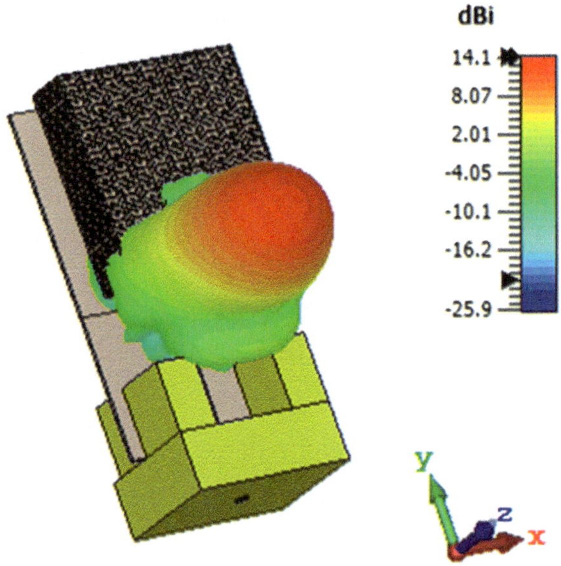


**Fig. 3.28** 3D radiation pattern of the patch antenna with MTM lens loading at 28.3 GHz ([23], Reprinted with permission from IEEE)

The poynting vector is plotted for cut plane $y = 0$ at $\omega t = 0$ for both the configurations, it is evident that both medias behave similarly to focus the radiated power along the z-axis when loaded at the same height from the patch. The 3D radiation pattern at 28.3 GHz of the patch with MTM lens is shown in Fig. 3.28, we can see a broadside radiation with peak gain 14.1 dBi and beamwidth of 27.4° and 36.7° is obtained in the E-and H-planes respectively.

The MTM realized lens is fabricated using the photolithography process and each layer is stacked over each other using 0.254 mm thick substrate at the corners of MTM layer supported using glue. Figure 3.29 shows the MTM lens of size $14.7 \times 14.86\ \text{mm}^2$ loaded on to a linearly polarized patch antenna for measuring the lens performance and the 3D printed stand which supports the MTM lens over the patch. The MTM lens is placed at $h = 3.5$ mm ($\sim 0.33\lambda_0$ at 28 GHz) from the MPA, reflection coefficient is measured using a Vector Network Analyzer (VNA). To verify the simulation results the prototype antenna is tested. Measured and simulated reflection coefficient is plotted in Fig. 3.30. The MTM lens loaded to the patch confirms the impedance matching and resonance at 28 GHz.

The measured radiation performance of the patch antenna with the MTM lens loading is compared with the simulated results. The radiation patterns in the H- and E-planes is plotted in Fig. 3.31. The normalized co-pol, $E_{co-pol}$ and $H_{co-pol}$ confirm radiation along the broadside direction with front-to-back ratio (FBR) greater than 20 dB when the MTM lens is loaded to the patch antenna. The cross-pol $E_{\times-pol}$ and $H_{\times-pol}$ along the z-direction are less than −40 dB and −35 dB respectively. The focusing of the fields along the z-directions due to the high and low epsilon values results in enhanced gain due to the MTM lens.

**Fig. 3.29** Photograph of the MPA with HLE lens realized using MTM layers

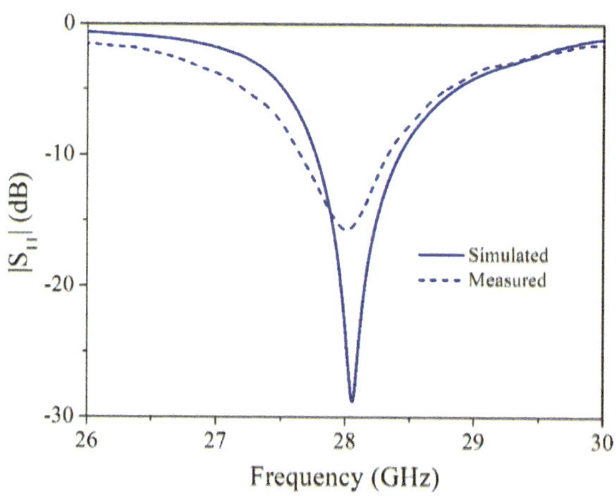


**Fig. 3.30** Measured and simulated input reflection coefficient of the patch antenna with the MTM lens loading ([23], Reprinted with permission from IEEE)

The measured gain of the antenna with MTM lens is compared with patch antenna gain in Fig. 3.32. Gain enhancement of about 7 dB is achieved with the MTM lens with measured peak gain of 13.8 dBi at 28 GHz in the broadside direction. The aperture efficiency ($e_a$) of the MPA with MTM lens having physical aperture area, $A_p$ can be calculated by, $e_a = \frac{10^{\left(\frac{G_{dB}}{10}\right)}\lambda^2}{4\pi A_p}$, where $G_{dB}$ is antenna gain in dB. The microstrip antenna with MTM lens has $A_p = 219\,\text{mm}^2$ and measured peak gain,

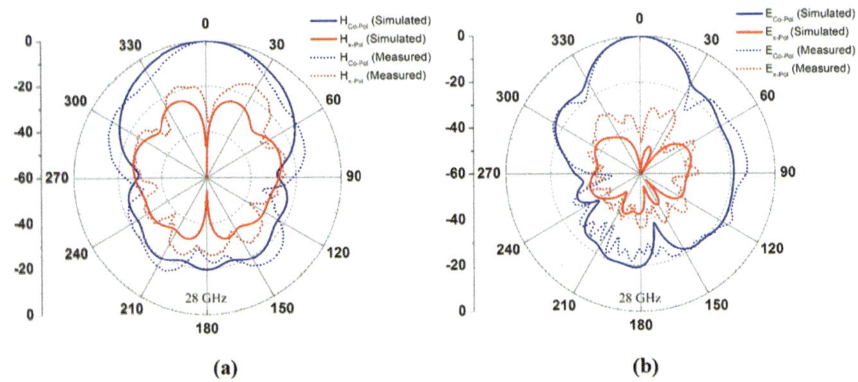


**Fig. 3.31** Measured and simulated radiation pattern **a** H-plane co- and cross-pol patterns, **b** E-plane co-pol and cross-pol patterns of the patch antenna with MTM lens at 28 GHz ([23], Reprinted with permission from IEEE)

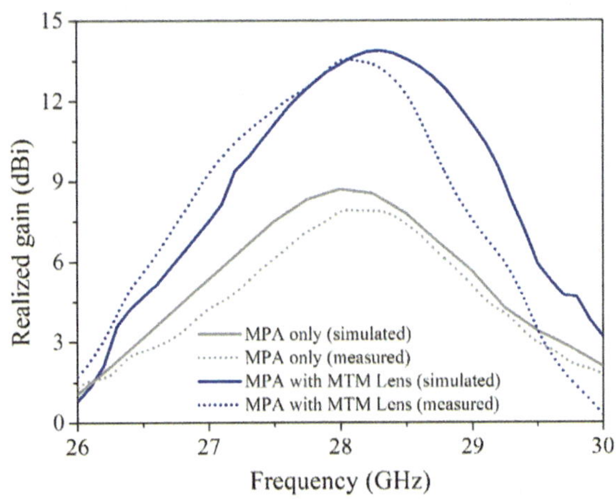


**Fig. 3.32** Measured and simulated realized gain comparison of the MPA with and without MTM lens ([23], Reprinted with permission from IEEE)

$G_{dB} = 13.8\,\text{dB}$; the aperture efficiency of the antenna comes out to be $e_a = 99.8\%$ at 28 GHz.

## 3.6 Conclusion

This Chapter introduces the need for high gain antennas for millimetre wave communication. Several techniques to enhance the gain and aperture efficiency of an antenna and their implementation issues at mmWaves are discussed in detail. All dielectric superstrate for gain enhancement is presented at 28 GHz, which results in 4.1 dB of gain enhancement. Later in the Chapter, a new gain enhancement technique using biaxial anisotropic slab is discussed in detail. The technique is implemented using metamaterials and results in an aperture efficiency of 99%.

## References


1. Balanis CA (2009) Antenna theory: analysis and design, 3rd edn. Wiley India Pvt. Limited [Online]. Available: https://books.google.co.in/books?id=2SJ0CgAAQBAJ
2. Attia H, Abdelghani ML, Denidni TA (2017) Wideband and high-gain millimeter-wave antenna based on FSS Fabry-Perot cavity. IEEE Trans Antennas Propag 65(10):5589–5594. https://doi.org/10.1109/TAP.2017.2742550
3. Ellis TJ, Rebeiz GM (1996) MM-wave tapered slot antennas on micromachined photonic bandgap dielectrics. In: 1996 IEEE MTT-S international microwave symposium digest, vol 2, pp 1157–1160. http://doi.org/10.1109/MWSYM.1996.511235
4. Singh AK, Abegaonkar MP, Koul SK (2017) High-gain and high-aperture-efficiency cavity resonator antenna using metamaterial superstrate. IEEE Antennas Wirel Propag Lett 16:2388–2391. https://doi.org/10.1109/LAWP.2017.2719864
5. Ge Y, Esselle KP, Bird TS (2012) The use of simple thin partially reflective surfaces with positive reflection phase gradients to design wideband, low-profile EBG resonator antennas. IEEE Trans Antennas Propag 60(2):743–750. https://doi.org/10.1109/TAP.2011.2173113
6. Hosseini A, Capolino F, Flaviis FD (2015) Gain enhancement of a V-band antenna using a Fabry-Pérot cavity with a self-sustained all-metal cap with FSS. IEEE Trans Antennas Propag 63(3):909–921. https://doi.org/10.1109/TAP.2014.2386358
7. Lalbakhsh A, Afzal MU, Esselle KP, Smith SL, Zeb BA (2019) Single-dielectric wideband partially reflecting surface with variable reflection components for realization of a compact high-gain resonant cavity antenna. IEEE Trans Antennas Propag 67(3):1916–1921. https://doi.org/10.1109/TAP.2019.2891232
8. Kim JH, Ahn C, Bang J (2016) Antenna gain enhancement using a holey superstrate. IEEE Trans Antennas Propag 64(3):1164–1167. https://doi.org/10.1109/TAP.2016.2518650
9. Pandey GK, Meshram MK (2015) Anisotropic artificial material with ENZ and high refractive index property for high gain Vivaldi antenna design. In: 2015 IEEE 15th Mediterranean microwave symposium (MMS), pp 1–4. http://doi.org/10.1109/MMS.2015.7375382
10. Lv YL et al (2013) A zero index metamaterial lens for gain enhancement of patch antenna and H-plane horn antenna. In: 2013 IEEE international wireless symposium (IWS), pp 1–4. http://doi.org/10.1109/IEEE-IWS.2013.6616849
11. Dadgarpour A, Zarghooni B, Virdee BS, Denidni TA (2016) One- and two-dimensional beam-switching antenna for millimeter-wave MIMO applications. IEEE Trans Antennas Propag 64(2):564–573. https://doi.org/10.1109/TAP.2015.2508478
12. Yuan LH, Tang WX, Li H, Cheng Q, Cui TJ (2014) Three-dimensional anisotropic zero-index lenses. IEEE Trans Antennas Propag 62(8):4135–4142. https://doi.org/10.1109/TAP.2014.2322898
13. Enoch S, Tayeb G, Sabouroux P, Guérin N, Vincent P (2002) A metamaterial for directive emission. Phys Rev Lett 89(21):213902. http://doi.org/10.1103/PhysRevLett.89.213902
14. Zhang S, Arya RK, Pandey S, Vardaxoglou Y, Whittow W, Mittra R (2016) 3D-printed planar graded index lenses. Antennas Propag IET Microw 10(13):1411–1419. https://doi.org/10.1049/iet-map.2016.0013
15. Afzal MU, Esselle KP, Zeb BA (2015) Dielectric phase-correcting structures for electromagnetic band gap resonator antennas. IEEE Trans Antennas Propag 63(8):3390–3399. https://doi.org/10.1109/TAP.2015.2438332
16. Afzal MU, Esselle KP (2016) A low-profile printed planar phase correcting surface to improve directive radiation characteristics of electromagnetic band gap resonator antennas. IEEE Trans Antennas Propag 64(1):276–280. https://doi.org/10.1109/TAP.2015.2493159

17. Lalbakhsh A, Afzal MU, Esselle KP, Smith SL (2019) Wideband near-field correction of a Fabry-Perot resonator antenna. IEEE Trans Antennas Propag 67(3):1975–1980. https://doi.org/10.1109/TAP.2019.2891230
18. Hayat T, Afzal MU, Lalbakhsh A, Esselle KP (2019) 3-D-printed phase-rectifying transparent superstrate for resonant-cavity antenna. IEEE Antennas Wirel Propag Lett 18(7):1400–1404. https://doi.org/10.1109/LAWP.2019.2917767
19. Wani Z, Abegaonkar MP, Koul SK (2019) Gain enhancement of a millimeter wave antenna using stacked dielectric-slabs superstrate. In: 2019 IEEE Indian conference on antennas and propogation (InCAP), pp 1–3. http://doi.org/10.1109/InCAP47789.2019.9134536
20. Edwards B, Alù A, Young ME, Silveirinha M, Engheta N (2008) Experimental verification of epsilon-near-zero metamaterial coupling and energy squeezing using a microwave waveguide. Phys Rev Lett 100(3):033903. http://doi.org/10.1103/PhysRevLett.100.033903
21. Kong JA (1986) Electromagnetic wave theory. Wiley [Online]. Available: https://books.google.co.in/books?id=OposAAAAYAAJ
22. Marcuse D (1972) Light transmission optics. Van Nostrand Reinhold, New York
23. Wani Z, Abegaonkar MP, Koul SK (2020) High-low-epsilon biaxial anisotropic lens for enhanced gain and aperture efficiency of a linearly polarized antenna. IEEE Trans Antennas Propag 68(12):8133–8138. https://doi.org/10.1109/TAP.2020.2983787
24. Orfanidis SJ (2016) Electromagnetic waves and antennas. https://www.ece.rutgers.edu/~orfanidi/ewa/. Accessed Feb 14, 2019

# Chapter 4
# Beam Tilting and Deflection Type Millimeter Wave Antennas

## 4.1 Introduction

The high gain antennas for mmWave communication discussed in the previous Chapter have narrow beamwidth which results in reduced multipath effects however, the beam alignment remains a challenge. The beam alignment and maintaining link between the transmitter and the receiver at mmWaves requires robust antennas whose radiation beam can be configured to a desired direction. The beam tilting antennas for the base stations usually use electrical or mechanical installations to tilt the radiation beam. The radiation beam tilting or deflection can be done electronically by having phase shifters integrated with the antenna elements in the array to steer the beam to a desired direction [1]. Other electronic techniques for beam tilting use RF MEMS and varactor diodes, however integrating these elements with antennas results in decreased gain due to losses in these elements [2].

Recently some interesting techniques using metamaterials to tilt/deflect the radiation beam from an antenna have been reported. The beam deflection is achieved by integrating the artificial media with tailored media properties with the antennas that results in compact and low loss beam tilting systems.

## 4.2 Beam Tilting Using Artificial High Permittivity Media

The end-fire antennas have become popular choice for mmWaves due to their low profile and ease of integration with the dielectrics/metamaterials. One of the basic approaches of achieving beam deflection can be done using off-axis dielectric lens in front of an antenna [3]. However, due to the large size of the dielectric lens the overall system becomes bulky for the next generation communication system. Recently researchers have developed novel approaches of tilting the radiation beam by using engineered materials such as artificial high permittivity media as in [4].



S. K. Koul and Z. Wani, *Novel Millimetre Wave Antennas for MIMO and 5G Applications*, Lecture Notes in Electrical Engineering 819,
https://doi.org/10.1007/978-981-16-7278-1_4

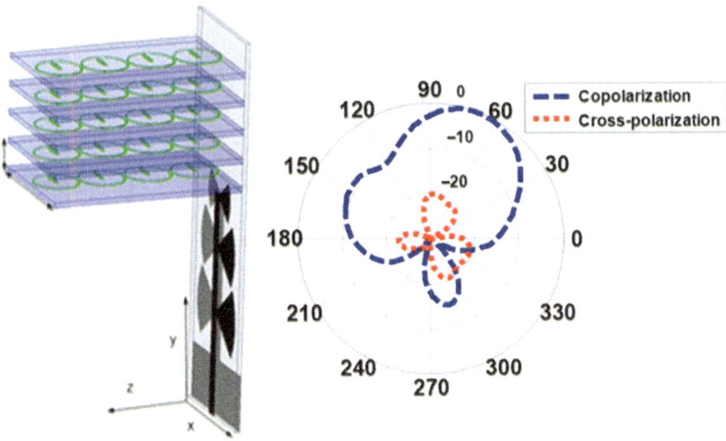


**Fig. 4.1** End-fire bow-tie antenna with artificial high dielectric material loading in H-plane ([4], Reprinted with permission from IEEE)

The beam deflection in the elevation plane is achieved by having artificial high permittivity media placed on top of the substrate on which the bowtie antenna is printed. The phase difference between EM waves from the substrate and the artificial media loaded vertically as shown in Fig. 4.1 results in tilted radiation beam towards the media. The artificial high dielectric loading is designed using double G-shaped resonators which provide a higher refractive index of 1.95 as compared with the substrate on which the antenna is printed. The antenna presented in Fig. 4.1 provides the beam deflection of 32° in the elevation plane from the end-fire direction.

The above concept of beam tilting can be used to achieve beam deflection in azimuthal plane using the high permittivity media in front of an end-fire antenna. To achieve the beam deflection, first a dipole antenna is designed at millimetre wave frequency band from 25 to 32 GHz. The dipole antenna is designed on the Neltec substrate having thickness 0.254 mm and dielectric constant of 2.2. The geometry of the dipole antenna is shown in Fig. 4.2 along with the input reflection coefficient and the E-plane radiation pattern of the antenna. This antenna covers the 28 GHz frequency band with $|S_{11}| < -10$ dB and has end-fire radiation properties with a gain of about 5.9 dBi. The dipole antenna is having 3-dB beam-width of 64.5° in the XOY-plane with the maximum radiation along the end-fire direction (along, $\phi = 90°$).

Now to tilt the radiation beam on left/right from the end-fire direction, a high permittivity media is placed in front of the antenna. The high permittivity media can be generated by modulating the effective permittivity of the substrate on which the antenna is designed. The change in the substrate permittivity is done by printing array of capacitively-loaded-loops (CLL) on one side of the substrate in front of the dipole antenna. The design of the CLL unit cell along with the extracted parameters is discussed below.

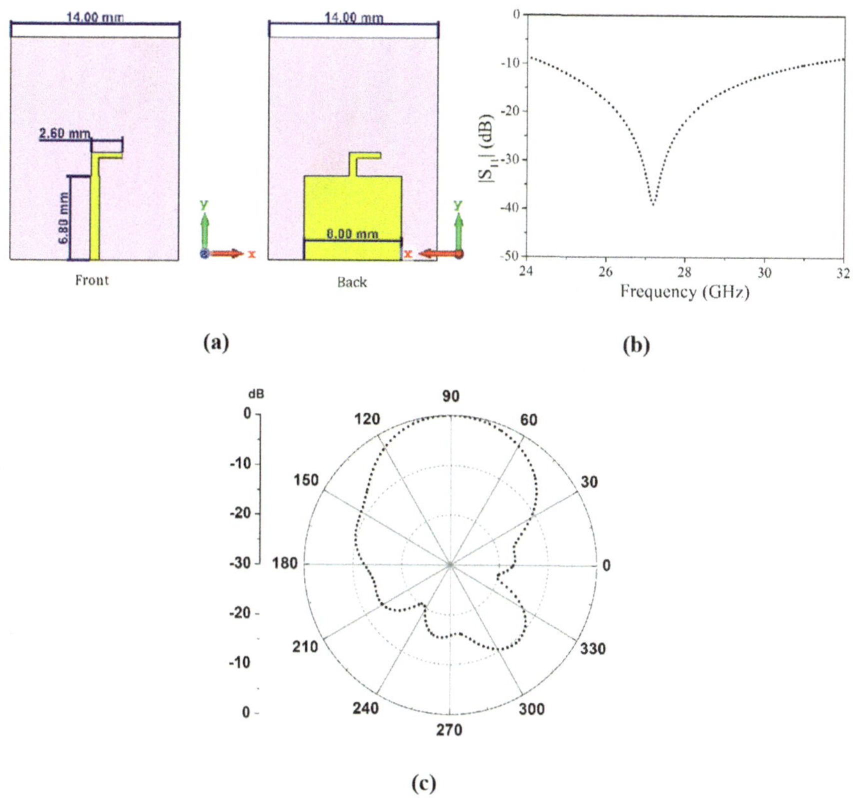


**Fig. 4.2** **a** Geometry of dipole antenna, **b** $|S_{11}|$ dB and **c** xoy-plane radiation pattern

### *4.2.1 Capacitively Loaded Loop Unit Cell*

The capacitively-loaded loop (CLL) metamaterial unit-cell structure is the simplified version of split ring resonators (SRR). The CLL unit-cell structure in [5] can be scaled at 28 GHz for high refractive index. The high refractive index at 28 GHz can be obtained by choosing the resonant frequency of the unit cell at higher frequencies. The electromagnetic waves from the dipole antenna have polarization along the x-direction, thus an electric resonant unit-cell is to be designed accordingly. This unit-cell can easily be integrated with the planar end-fire antenna and therefore was chosen. The unit-cell shown in Fig. 4.3a, is designed on a Neltec NY9220 substrate with height, $h_{sub} = 0.254$ mm. The unit-cell was designed and simulated using CST Studio Suite with perfect electric and magnetic boundary conditions applied to yz- and xy-planes respectively. The unit-cell is excited by a plane wave along the y-direction with electric polarization along x-axis as shown in the simulation setup given in Fig. 4.3a. The simulated S-parameters of the unit-cell for the plane wave excitation are shown in Fig. 4.3b, the plot shows $|S_{11}|$ and $|S_{21}|$ as a function

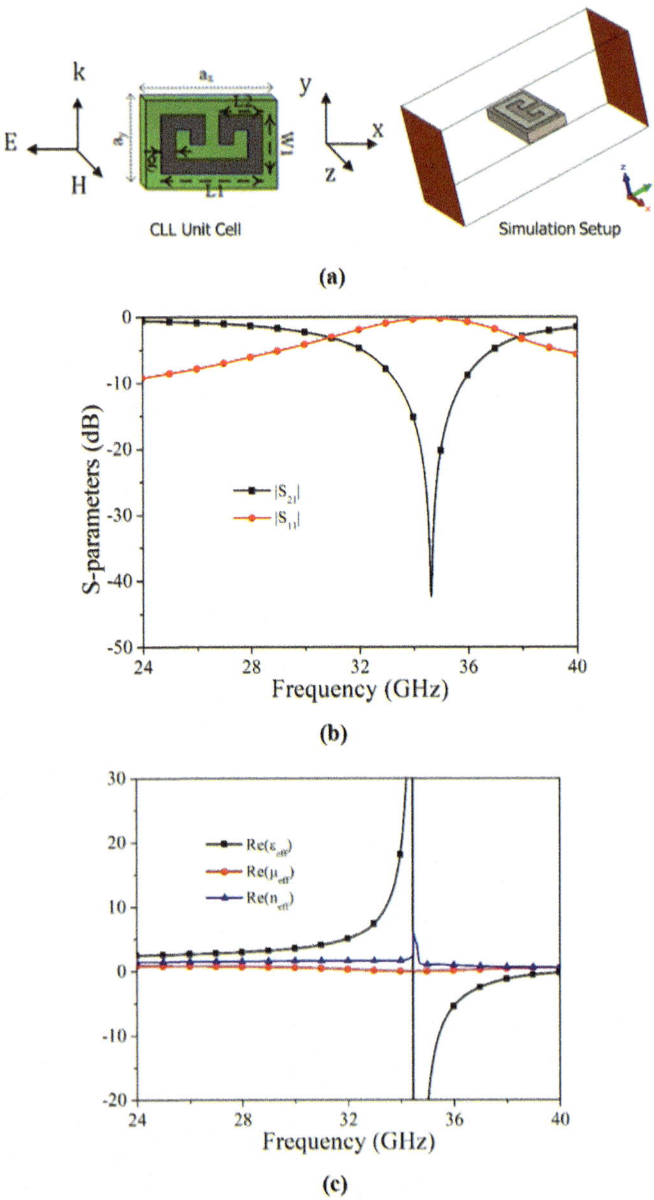


**Fig. 4.3** **a** Geometrical description of the unit cell ($a_x = 1.64$ mm, $a_y = 1$ mm, L1 = 1.4 mm, L2 = 0.6 mm, W1 = 0.8 mm, g = 0.2 mm), **b** S-parameters and **c** retrieved real part of refractive index ($Re(n_{eff})$), effective permittivity ($Re(\epsilon_{eff})$) and permeability ($Re(\mu_{eff})$) from simulated S-parameters ([7], Reprinted with permission from Wiley)

of frequency. As observed, $|S_{21}|$ resonance dip occurs around 35 GHz which can be tuned by the length and capacitive gap width of the CLL unit cell.

The CLL unit-cell parameters such as, permittivity, permeability and refractive index are extracted using the algorithm described in [6] from the simulated S-parameters. This algorithm retrieves the effective properties of the metamaterial slab from the S-parameters when excited with a plane wave at normal incidence. This improved retrieval method is very effective even when the magnitude of the S-parameters is small. The retrieved effective parameters of the CLL unit are plotted in Fig. 4.3c. As observed from Fig. 4.3c, the unit cell is having high permittivity and low permeability across the frequency range 25–30 GHz. Thus, choosing resonance at higher frequency of 35 GHz results in positive higher permittivity value in the lower frequency band. The extracted refractive index $n_m$ of the designed CLL unit cell is between 1.9 and 2.1 in the frequency range 25–30 GHz. The dipole antenna gain can be enhanced by printing the array of unit cell structure on the substrate which results in higher effective refractive index than the refractive index of the substrate, $n_{sub} = 1.48$.

### 4.2.2 *Design and Mechanism of Beam Deflection*

The beam deflection is achieved by printing unit cell array at an offset position along x-axis as shown in Fig. 4.4a. To verify the radiation beam tilt using the offset loading of CLL metamaterial unit cell array on the dipole antenna, a prototype antenna is fabricated. The photograph of the antenna is shown in Fig. 4.4b. This prototype is characterized using the vector network analyser and the measured $|S_{11}|$ (dB) is plotted with the simulated input reflection coefficient in Fig. 4.4c. The offset metamaterial loading to dipole antenna is like placing two different media with different refractive indices (low and high) in front of an electromagnetic (EM) point source to create propagation phase difference of the EM rays travelling through low and high refractive index media, thereby the total radiated power can be directed along a different direction. The mechanism of the beam tilting is depicted in Fig. 4.5a. The tilt angle is $(90 - \Phi)$, where $\Phi$ the angle from the x-axis, is dependent upon the path difference between the path length 'BC + Ls' of EM rays travelling through the substrate and the path length 'Lm' of EM rays travelling through the high refractive index medium [8]. The titling angle can be calculated by equating the path lengths of the EM rays in the two media using:

$$n_{sub} L_s + W_m \cos\phi = n_m L_m \tag{4.1}$$

The beam pointing angle from the x-axis is given as:

$$\cos\phi = \frac{(n_m - n_{sub}) L_m}{W_m} \tag{4.2}$$

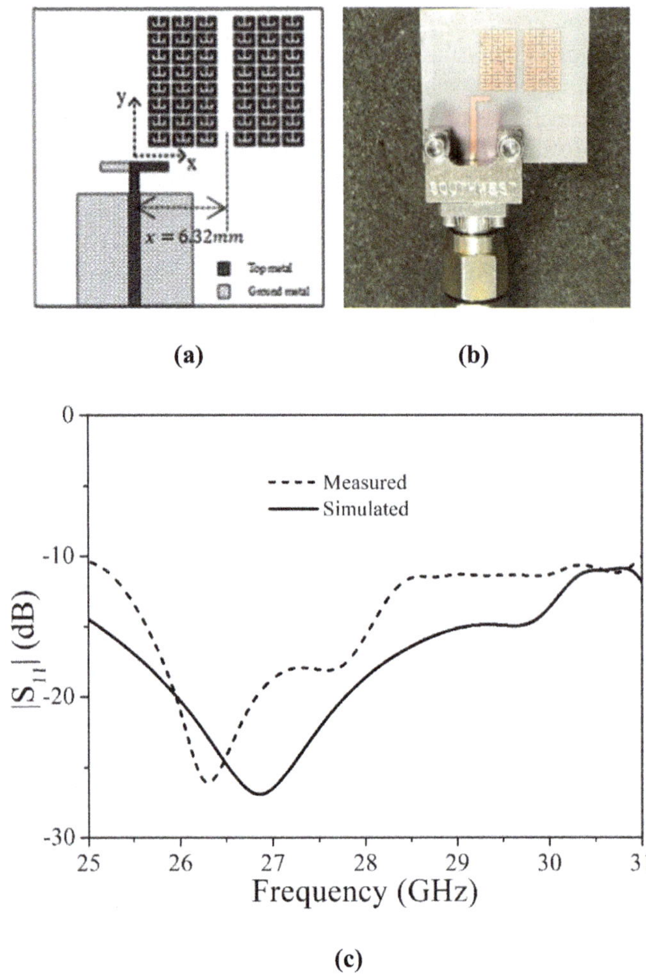


**Fig. 4.4** **a** Dipole with offset metamaterial loading (w/ML-offset), **b** photograph of the fabricated prototype and **c** reflection coefficient of the antenna with offset loading ([7], Reprinted with permission from Wiley)

Since $n_m > n_{sub}$, the tilting angle will depend upon the difference in the refractive indices of the two media. Figure 4.5b shows the radiation pattern of the dipole antenna without ML, when the metamaterial unit cells are loaded at an offset, $x = +6.32$ mm, $x = +2.32$ mm and $x = -6.32$ mm with respect to symmetrical loading of the antenna array. The radiated beams deflect towards the array of metamaterial unit cells, results in bending of EM beam to the right or left side. The loading of metamaterial at an offset of $x = -6.32$ mm, $x = +2.32$ mm and $x = +6.32$ mm, tilts the radiation beam to $\Phi = \mathbf{115°}$, **80°** and **65°** from **90°** (symmetrical loading case), resulting in bending of EM rays by **−25°**, **10°** and **25°** to the left and right, respectively. The

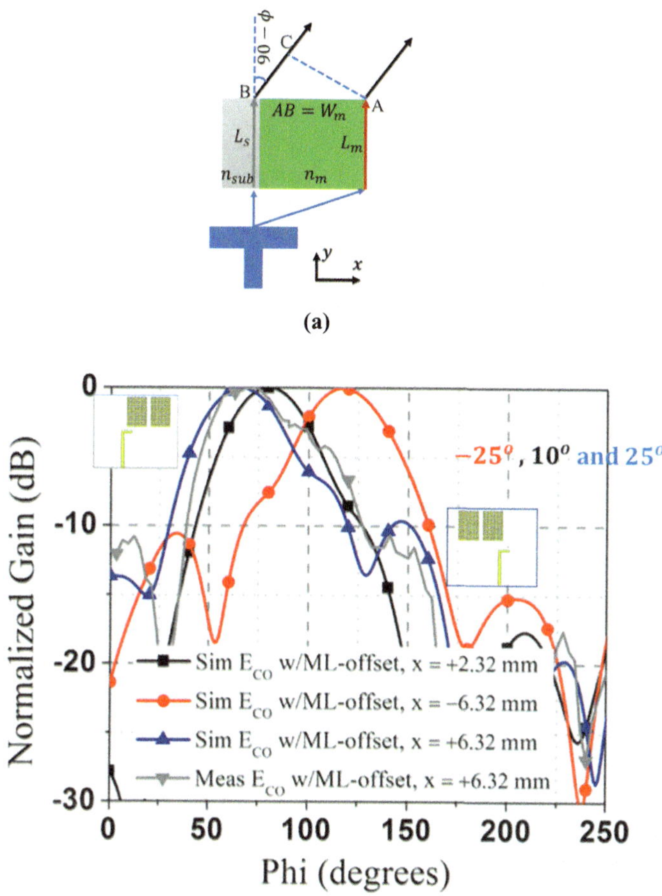


(a)

(b)

**Fig. 4.5** **a** Mechanism of beam tilting, **b** measured and predicted normalized gain pattern in xoy plane at 28 GHz without ML (wo/ML), with offset metamaterial loading (ML) in +x and −x directions ([7], Reprinted with permission from Wiley)

amount of tilt in the radiation beam is directly proportional to the offset printing of the metamaterial cells on the right and left sides. The simulation results are verified using the (4.1) and (4.2) by inputting the dimensions of the high refractive index surface. Now for an offset $x = +6.32$, the $L_m = 7$ mm and $W_m = 10.4$ mm, the calculated value of beam pointing angle for the offset comes out to be: $\phi = 66°$ and Tilting Angle $= 24°$.

The measured radiation pattern of the antenna for offset $x = +6.32$ mm confirms the beam tilting to $\Phi = \mathbf{65}°$ and is plotted in the Fig. 4.8b which further validates the tilting mechanism. The beam tilting effect can be studied from the profile of the radiated electric fields through high index surface. The electric field magnitude

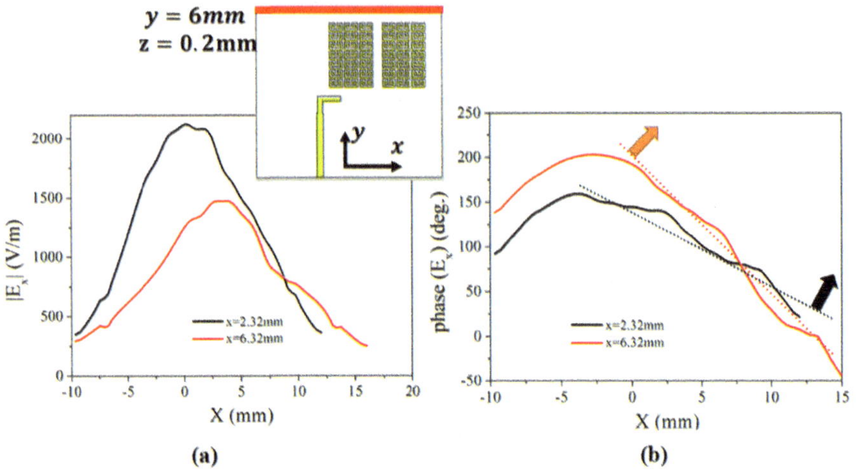


**Fig. 4.6** **a** Magnitude of the electric field along the red line, **b** phase of the electric field through the CLL array

and phase is plotted along x-axis at the top of the CLL array as shown in Fig. 4.6 for the offset position ($x = +6.32$ and $+ 2.32$ mm). The maximum magnitude of the electric field through the region shifts towards the high index region with offset position and thus radiation beam is tilted more with more offset position. Also, from the phase profile of the electric field, we can see that the slope of the fields through the CLL array increases with the offset loading which results in the power flow along different directions.

## 4.3 Huygens' Metasurface for Beam Deflection

The Huygens' metasurface consists of array of two-dimensional co-located electric and magnetic resonators to control the electromagnetic fields through the surface [9]. With the help of these metasurfaces, discontinuities can be imparted to the electromagnetic wave front which helps in shaping the wave front of the outgoing EM waves. The phase or the polarization of the propagating EM waves is controlled by array of sub wavelength resonators which introduce spatially varying phase/polarization. The Huygens' metasurface can be used for beam deflection by introducing phase shift across the surface using the phase delay resonators. These phase delay elements act as secondary sources and the design of these resonators is very crucial to have efficient metasurface.

In [10], first experimentally tested Huygens' metasurface is designed for bending of optical beam at 1.5 μm. The metasurface consists of array of isotropic unit cells to provide both electric and magnetic response necessary for a Huygens' surface. The unit cell provides large phase coverage at a high transmission through the cell.

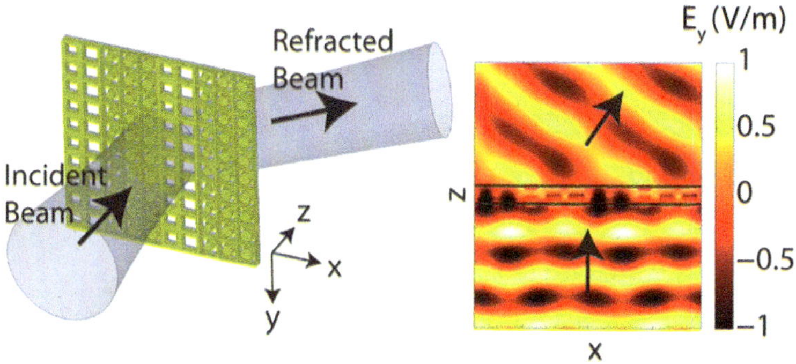


**Fig. 4.7** Thin Huygens's metasurface to refract incident beam with electric field profile depicting the bending of the incident beam ([10], Reprinted with permission from ACS)

To refract the normal incident beam, the surface consists of array of the unit cells with linear phase progression across the surface to bend the beam as shown in the Fig. 4.7. The designed metasurface efficiently bends the incident beam by 35° from the normal.

The concept of Huygens' metasurface can be used to antennas at millimetre frequencies to achieve tilting of the radiation beam. The beam tilting can be achieved by having array of transmission phase delay elements on a surface placed in front of an antenna.

### *4.3.1 Design of Phase Delay Unit Cells*

In order to design a unit cell for the Huygens' metasurface, the unit cell should provide a high transmission with large phase coverage. The transmission type unit cell reconfigures the phase of the propagating wave and is designed using a double layered substrate ($\epsilon_r = 2.2$, $h = 20$ mil) separated with an air gap of 1 mm as shown in Fig. 4.8a. The phase of the propagating wave is controlled by the total perimeter of the rectangular loop etched on the metallic sheet of each substrate. The unit cell is studied using FEM simulation with unit cell boundaries on each side in the xy-plane. The simulated $|S_{21}|$ and transmission phase is plotted in Fig. 4.8b for different values of $a$ (loop length). The plot shows the linear transmission phase variations of the wave with the parameter $a$ (in mm) of the unit cell. By varying the parameter value from 2.2 to 2.82 mm, a phase delay of 232° can be achieved with a very high transmission above −5 dB through the unit cell. The unit cell is further studied for different polarizations in the simulations. Figure 4.9 shows the transmission and reflection coefficient of the unit cell for TE (mode 1) and TM (mode 2) polarizations. It is clear that the response of the unit cell is insensitive to mode of excitation as the response is unchanged for two polarizations of the excitation.

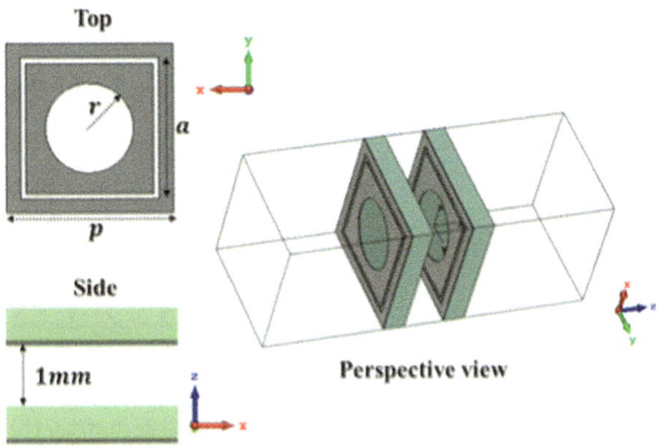


(a)

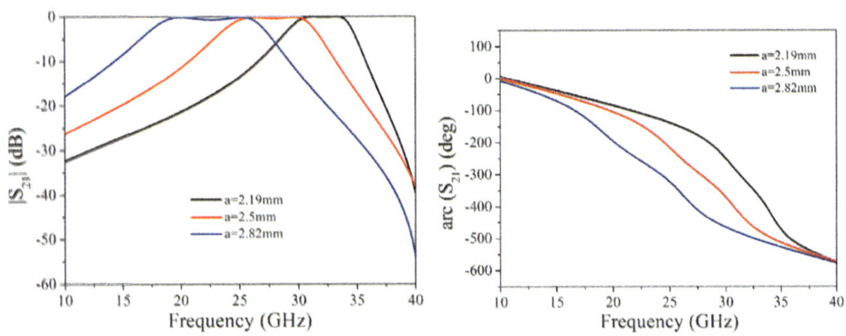


(b)

**Fig. 4.8** **a** Geometry of the unit cell, **b** simulated transmission magnitude and phase of the unit cell

## *4.3.2 Design of Huygens' Metasurface*

The beam deflection metasurface is realized by having linear phase progression across the surface using the phase delaying metamaterial unit cells. These unit cell resonators act as secondary sources and the required beam tilt can be achieved by tuning the phase delays at each unit cell. The phase delays at each cell location can controlled by the physical loop length of the unit cell. With the help of the array theory, the beam pointing angle ($\theta_s$) is given as:

$$\sin\theta_s = \frac{\lambda}{2\pi d}\Delta\phi \tag{4.3}$$

where $\Delta\phi$ is the phase delay between the two successive unit cells and $d$ is the distance between the two adjacent unit cells. The phase shift between unit cells can be written from the (4.3) as:

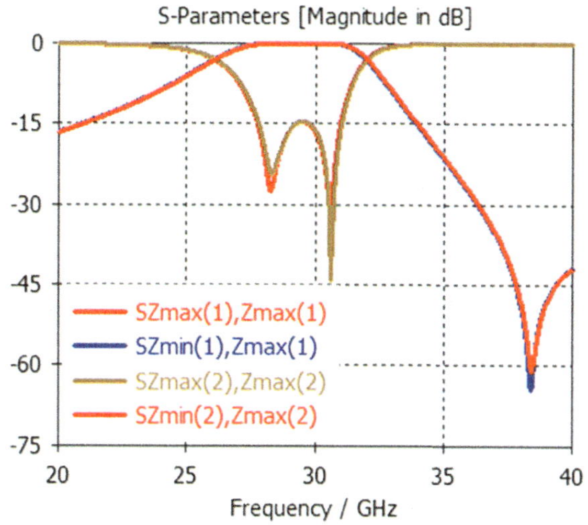


**Fig. 4.9** Transmission and reflection response of the unit cell for TE mode (1) and TM mode (2)

$$\Delta\phi = \frac{2\pi . d . \sin\theta_s}{\lambda} \tag{4.4}$$

With the help of (4.4), the Huygens' metasurface (HMS) can be designed for beam deflection of $\theta_s$ from the normal. In order to have a beam tilt towards the x-axis, six types of unit cells having progressive phase delay of $\Delta\phi$ are arranged from left to right as seen from Fig. 4.10. Each unit cell (cell1 to cell6) provides an additional phase delay of $\Delta\phi$ with respect to the cell on the left resulting in bending of the beam towards x-axis. For a beam tilt of 30°, the required phase delay between the cells is about 55° with a separation of 3 mm between the cells.

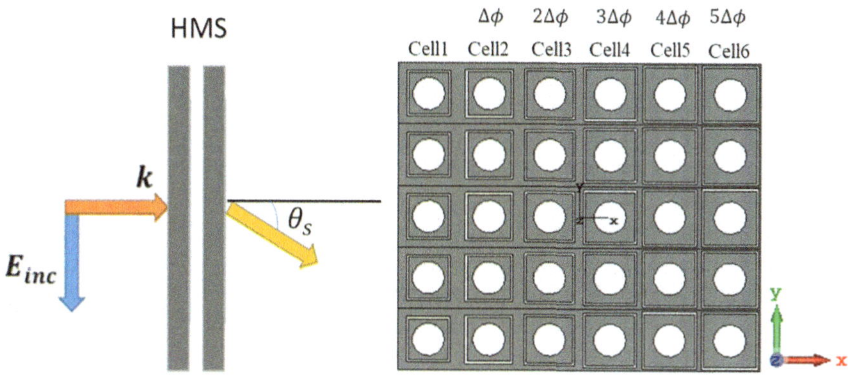


**Fig. 4.10** Configuration of the HMS for beam tilting

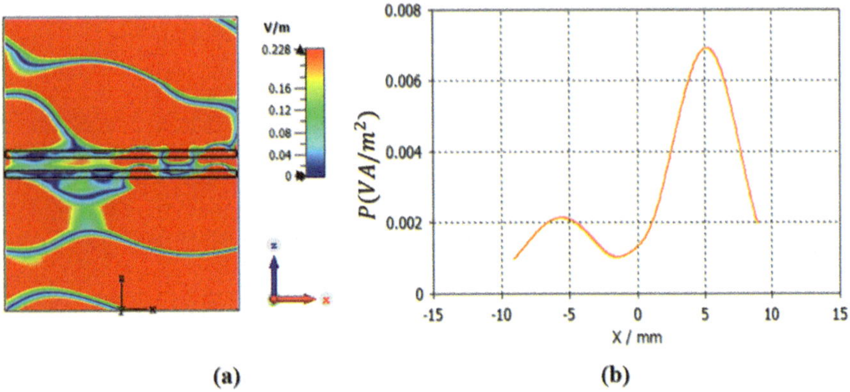


**Fig. 4.11** **a** Snapshot of the electric field through the HMS and **b** power flow along x-axis

The metasurface refraction properties are studied using simulation tools with normal incidence of plane wave. The performance of the designed Huygens' metasurface is shown in Fig. 4.11. Figure 4.11a shows the snapshot of the electric field through the metasurface when a plane wave is incident normally at 28 GHz. From the electric field plot, we can see that the beam is refracted towards the x-axis by 30° with $\Delta\phi = 55°$. Further the power flow through the HMS can be plotted along the x-axis (y = 0) at z = 5 mm above the surface. The plot shows that the maximum power density is at an offset position from the normal which results in beam tilting of 30°.

## 4.4 Microstrip Patch Antenna with the Huygens' Metasurface

The designed metasurface is then placed above a microstrip patch antenna designed at 28 GHz on a Rogers 5880 substrate. The performance of the system is next evaluated with the help of simulations as shown in the Fig. 4.12a. As observed, the loading of the HMS above the patch antenna does not disturb the port impedance of the antenna due to the high transmission property of the metasurface. The input reflection coefficient of the antenna with the HMS is shown in the Fig. 4.12b, the plot shows that the HMS does not affect the resonant frequency of the patch antenna. The 3D radiation pattern of the patch antenna with the HMS is shown in Fig. 4.13a at 28 GHz. Figure 4.13b shows that the radiation from the antenna is directed away from the z-axis by angle of 30° due to the HMS loading at a height of 5.3 mm from the patch antenna. The metasurface also enhances the gain of the antenna by directing most of the power along a single direction. The patch antenna with the HMS offers about 11.5 dB gain which is quite useful at millimetre wave frequencies.

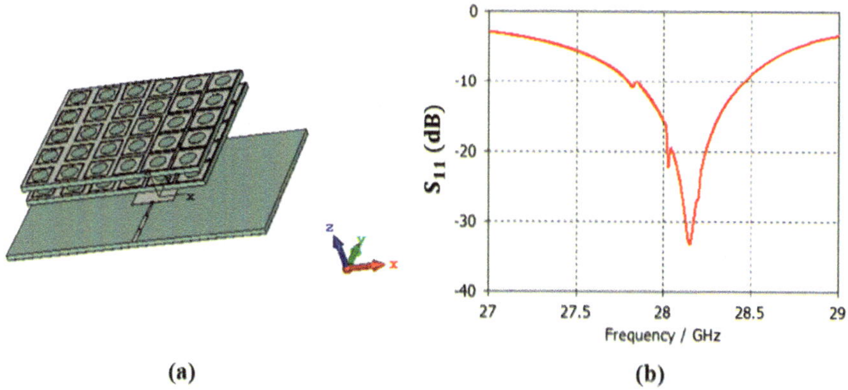


**Fig. 4.12** **a** Configuration of the antenna with HMS, **b** input reflection coefficient of the antenna with HMS

## 4.5 Dipole Antenna with the Huygens' Metasurface

Now in order to achieve a beam deflection of 20° in the azimuthal plane, the designed Huygens' metasurface is tweaked using (4.4). With $\Delta\phi = 34.5°$, required beam tilt of 20° can be achieved. This is done by loading the modified HMS in front of a dipole antenna as shown in Fig. 4.14. The dipole antenna covers the frequency band 27–30 GHz with HMS, the input reflection of the dipole antenna with the HMS is below −10 dB over this band.

The radiation performance of the end-fire antenna with HMS for different frequency samples is shown in Fig. 4.15. The 3D radiation pattern shows the tilting of the radiation beam by 20° at 27, 28 and 29 GHz. The antenna has a stable radiation performance with an average gain of about 10dBi throughout the frequency band.

## 4.6 Conclusion

This Chapter introduces different beam tilting techniques and implementation for millimetre wave communication. The techniques involving manipulation of the radiated fields using the metamaterials to achieve beam deflection are discussed in detail. The beam tilting technique using high refractive index to modulate the surface wave velocity using capacitively loaded loops is also presented. This technique has very low profile and can be implemented with any end-fire antenna on a planar substrate. Later in the Chapter, beam tilting using Huygens' metasurface is discussed which offer great deal of reconfigurability in terms of beam tilting angles. The implementation of this technique is carried out using very tiny phase gradient resonators arrayed together to have successive phase progression to modulate the wave-front of the outgoing wave. The metasurface is then loaded to the patch and dipole antenna

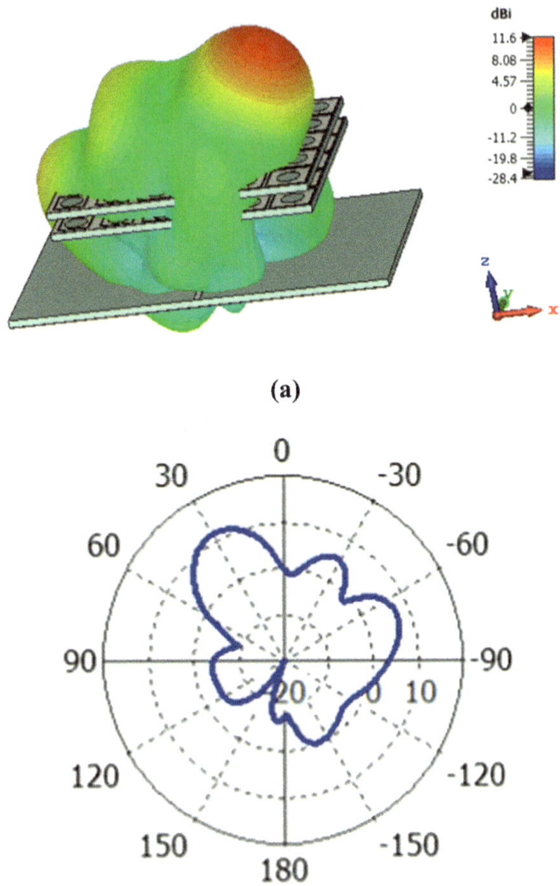


(b)

**Fig. 4.13** **a** 3D radiation pattern of the antenna and **b** H-plane radiation pattern at 28 GHz

to achieve titled radiation beam with respect to broadside and end-fire directions, respectively.

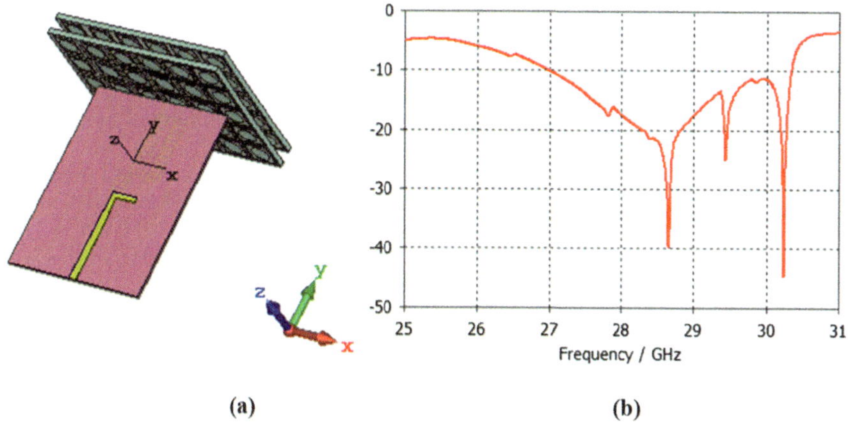


**Fig. 4.14** **a** Dipole antenna with HMS and **b** input reflection of the antenna

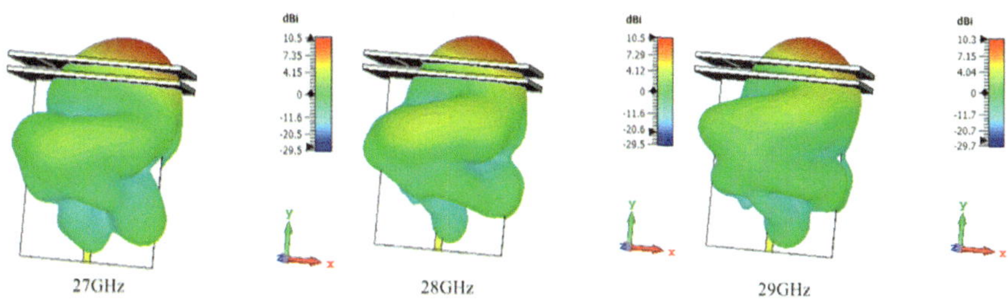


**Fig. 4.15** 3D radiation pattern of the dipole antenna with HMS at 27, 28 and 29 GHz

# References


1. Rajo-Iglesias E, Quevedo-Teruel O (2013) Tilting radiation patterns in linear arrays without phase shifters. IEEE Trans Antennas Propag 61(6):3360–3364. http://doi.org/10.1109/TAP.2013.2247552
2. Luther JJ, Ebadi S, Gong X (2012) A microstrip patch electronically steerable parasitic array radiator (ESPAR) antenna with reactance-tuned coupling and maintained resonance. IEEE Trans Antennas Propag 60(4):1803–1813. https://doi.org/10.1109/TAP.2012.2186265
3. Filipovic DF, Gauthier GP, Raman S, Rebeiz GM (1997) Off-axis properties of silicon and quartz dielectric lens antennas. IEEE Trans Antennas Propag 45(5):760–766. https://doi.org/10.1109/8.575618
4. Dadgarpour A, Zarghooni B, Virdee BS, Denidni TA (2015) Enhancement of tilted beam in elevation planc for planar end-fire antennas using artificial dielectric medium. IEEE Trans Antennas Propag 63(10):4540–4545. https://doi.org/10.1109/TAP.2015.2456937
5. Erentok A, Luljak PL, Ziolkowski RW (2005) Characterization of a volumetric metamaterial realization of an artificial magnetic conductor for antenna applications. IEEE Trans Antennas Propag 53(1):160–172. https://doi.org/10.1109/TAP.2004.840534
6. Chen X, Grzegorczyk TM, Wu B-I, Pacheco J, Kong JA (2004) Robust method to retrieve the constitutive effective parameters of metamaterials. Phys Rev E 70(1):01660. http://doi.org/10.1103/PhysRevE.70.016608

7. Wani Z, Abegaonkar MP, Koul SK (2019) Millimeter-wave antenna with wide-scan angle radiation characteristics for MIMO applications. Int J RF Microw Comput Aided Eng 29:e21564. https://doi.org/10.1002/mmce.21564
8. Dadgarpour A, Zarghooni B, Virdee BS, Denidni TA (2014) Beam tilting antenna using integrated metamaterial loading. IEEE Trans Antennas Propag 62(5):2874–2879. https://doi.org/10.1109/TAP.2014.2308516
9. Pfeiffer C, Grbic A (2013) Metamaterial Huygens' surfaces: tailoring wave fronts with reflectionless sheets. Phys Rev Lett 110(19). https://doi.org/10.1103/PhysRevLett.110.197401
10. Pfeiffer C, Emani NK, Shaltout AM, Boltasseva A, Shalaev VM, Grbic A (2014) Efficient light bending with isotropic metamaterial Huygens' surfaces. Nano Lett 14(5):2491–2497. https://doi.org/10.1021/nl5001746

# Chapter 5
# Beam Switching and Wide-Scan Antenna Array

## 5.1 Introduction

The previous Chapter describes how the beam tilting antennas with fixed radiation beam helps to maintain the line-of-sight wireless link. However, millimetre wave indoor communication is not necessarily LOS but can be both LOS and non-LOS and to maintain the communication link requires a smart antenna system having beam switching capabilities [1]. The beam switching antennas at millimetre wave frequencies usually consist of number of high gain antennas with fixed radiation beams. The radiation beams can be switched to a direction to avoid any signal blockage resulting due to objects in between the transmitter and the receiver. The beam switching antenna system does not provide optimum signal-to-interference ratio as the direction of arrival of the signal is not analysed. However the beam switching antennas with wide angular coverage provide considerable performance improvement at lower complexity and power consumption [2].

The technique of beam switching evolved from the curved reflector antennas which serves an area to which the antenna is pointing. However, each antenna with single beam serves a single angular area which is not economical in terms of use of space. To achieve higher efficiency and efficient use of space, switched beam antenna arrays using Butler matrix to have multiple beam from a single aperture are proposed in [3, 4].

## 5.2 Beam Switching Antenna Array

Traditionally the process of synthesizing radiation beams from an array of equally spaced antennas depends on the adjustable phase and amplitude distribution. The phase shifts at each antenna element helps in steering the beam to a given angle and amplitude distribution helps in shaping the pattern. In order to design a two-port



S. K. Koul and Z. Wani, *Novel Millimetre Wave Antennas for MIMO and 5G Applications*, Lecture Notes in Electrical Engineering 819,
https://doi.org/10.1007/978-981-16-7278-1_5

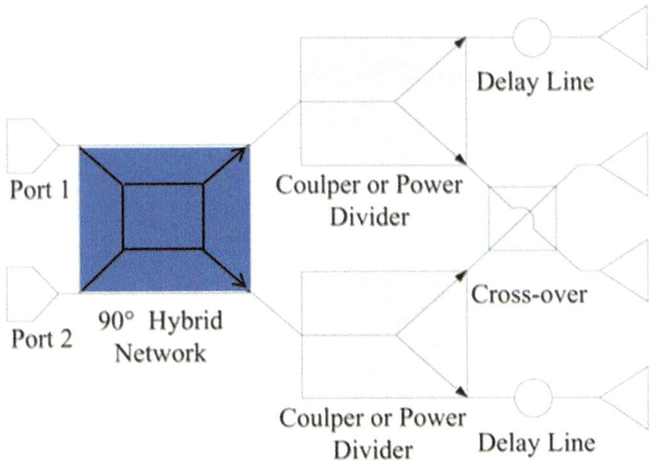


**Fig. 5.1** Feed network for two-port beam switching antenna

beam switching antenna with four antenna elements, a 2 × 4 feeding structure is to be implemented as shown in Fig. 5.1 [5]. The feeding network consists of branch line coupler with power dividers and a crossover to distribute the phase among the four antenna elements. For equally spaced antennas and delay lines 1 and 2 providing phase shifts of 0 and 40°, respectively, the antenna array system with the feeding network generates two radiation patterns along ±22° as shown in Fig. 5.2. Here each antenna element is basically a simple patch antenna designed at 28 GHz using the CST studio suite.

Based on the feed network configuration described in Fig. 5.1, a beam switching antenna is designed in [5]. The prototype antenna consists of an array of four identical antennas as shown in Fig. 5.3a. The feed network consists of hybrid coupler with two power dividers/coupler for phase distribution to four antenna ports by switching two input ports. The antenna beam can be switched to two different directions depending upon which port is excited as can be seen from the radiation patterns of the antenna when port 1 and port 2 is excited as shown in the Fig. 5.3b. The radiation beams are directed along ±22.5° with normalized side lobe level of −12 dB.

The Butler matrix used above for beam switching function consists of 3 dB hybrid couplers and power dividers to achieve fixed phase shifts. The Butler matrix can be extended to $2^n$ inputs and output beams as shown in Fig. 5.4. As the operation of these hybrid couplers is frequency dependent which restricts the use of this technique to narrow bandwidths only. Also beam shaping control is difficult using the Butler matrix as the inputs and outputs of the hybrid couplers are limited to two (or power of 2). Also, implementation of 8-way Butler matrix at mmWaves can result in low efficiency due to the huge feed losses.

At millimetre wave frequencies, beam switching can be implemented using SPDT/SP3T switches together with corporate feed network. As the switches at mmWaves are less costly than phase shifters, therefore beam switching array having

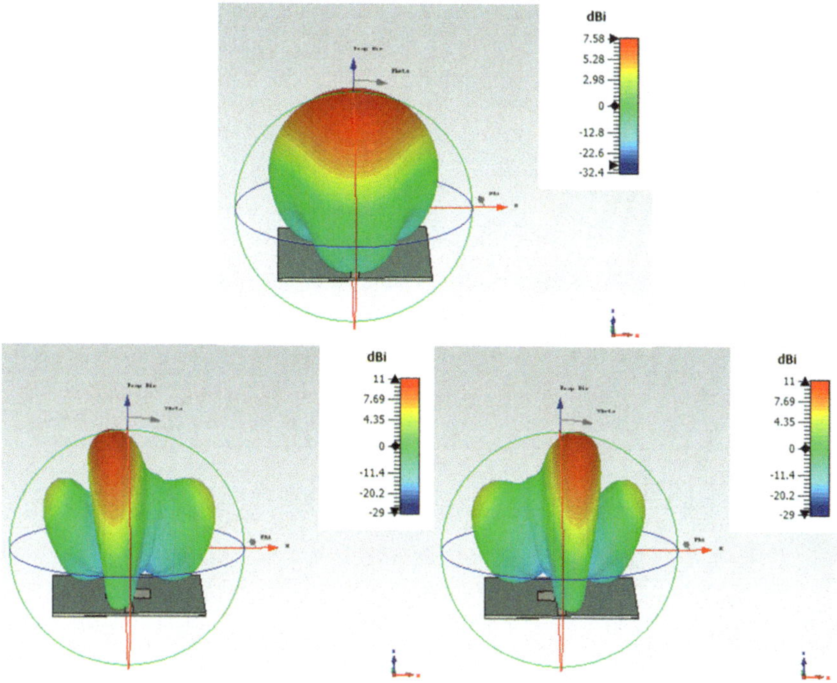


**Fig. 5.2** Radiation pattern of patch antenna and array of equally spaced four element array

2–3 output beams can be implemented at very low costs. A conformal beam switching antenna operating at 61 GHz is implemented using an SP3T switch in [6]. Three radiation beams are generated from an array of three antennas with each antenna consisting of 16 patch antennas as shown in Fig. 5.5a.

The single-pole three-throw (SP3T) switch operates in the range 60–90 GHz and the radiation beam can be switched by applying voltage to the control pins. The antenna generates three radiation beams directed along 0° and ±30° when the antenna is conformed on a Styrofoam cylinder. The measured radiation patterns from antennas 1 and 3 are shown in Fig. 5.5b.

## 5.3 Integrated Lens Antennas

The integration of lens with an array of planar antennas results in improved radiation characteristics. The dielectric/metamaterial lens can be driven by number of antennas placed at the focal plane of the lens [7]. The schematic of the lens with the antenna array is shown in Fig. 5.6. The radiation beams from the lens are directed along different directions based on which antenna at the focal plane is switched and

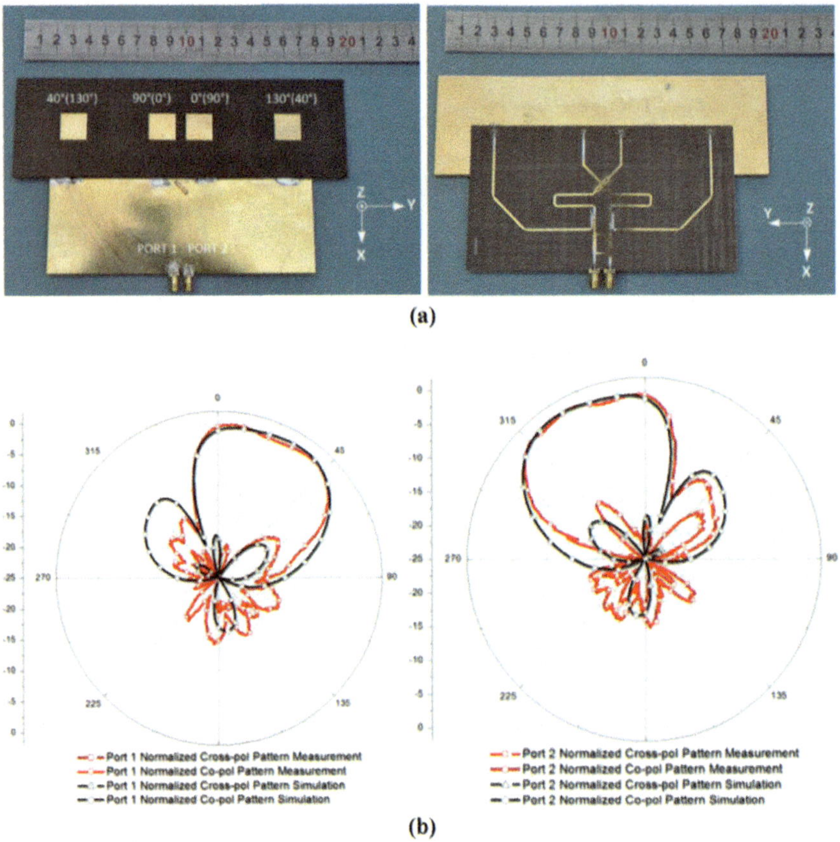


**Fig. 5.3** **a** Two port beam switching antenna and **b** 2D radiation pattern when port 1 and 2 are excited ([5], Reprinted with permission from IEEE)

activated [8]. The integration of a lens with an antenna array results in a highly directional radiation beam with lesser number of components as compared to phased arrays. Also by employing an RF/MEMS switch to activate the antennas offers lower losses and lower costs when compared to beam switching antennas using a Butler matrix [9].

A mmWave beam switching antenna consisting of four aperture coupled microstrip patch antennas with a quartz hemispherical lens operating at 60 GHz frequency band is presented in [10]. The quartz lens with radii of 7.5 mm and 12 mm is mounted onto the array of antennas as shown in Fig. 5.7a. Three SPDT RF switches, each switch offering 2.5 dB insertion loss, are mounted on the substrate to switch one of the four antennas. The measurement results confirm the electronic beam switching to ±35° and ±22° with the lens as shown in Fig. 5.7b. The maximum achieved gains are 18.4 dBi and 23.2 dBi with 7.5 mm and 12 mm radius lens respectively. Similarly in [11], a three-element slot antenna array with spherical dielectric lens to increase

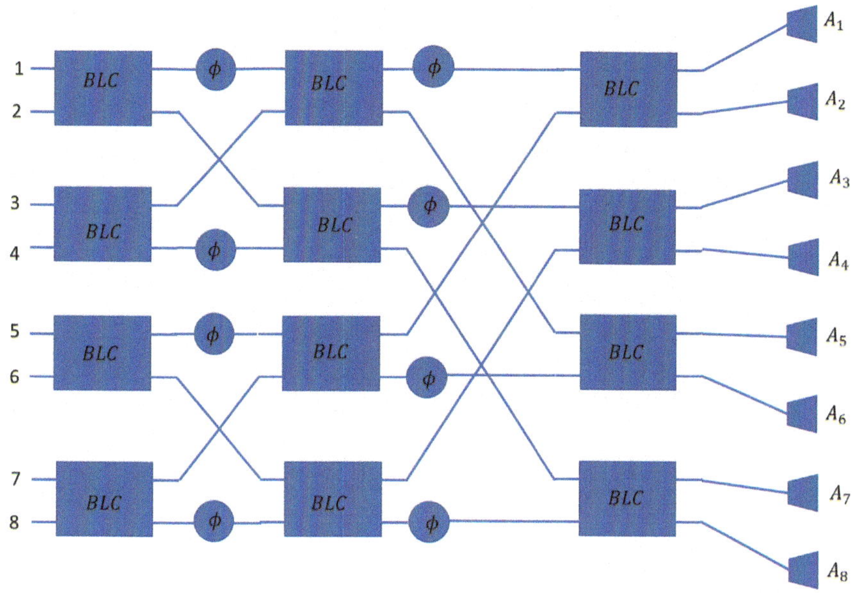


**Fig. 5.4** A butler matrix feed network for beam-switching

the beam scan and gain enhancement operating at 60 GHz is presented for end-fire applications. The measured results recorded radiation along 60°, 90° and 120° with a peak gain of 20 dB. This antenna is quite bulky with a volume of $11\lambda_o^3$ including the lens. To achieve large beam scanning with the above technique requires larger offset between the antenna elements. However increasing the distance between the antennas results in lower focusing with larger gain scan loss [8]. The gain of the antenna element decreases with the offset distance from the focal point of the lens and gain scan loss is the difference in the gains of the offset element with respect to the central element [12]. The concept of integrated lens antenna can be extended to 2D beam steering by employing 2D array of antenna elements with a lens. The integrated lens antennas have limitations in terms of only predefined angular spaces that can be covered using the antenna system and loss of the signal if the receiver antenna gets displaced from the location. Also use of dielectric lens with antenna array results in a bulky antenna system not suitable for mmWave handheld applications. Recently, the engineered materials such as metamaterials have replaced conventional dielectric lens. The design of metamaterial-based beam switching antenna is discussed in Sect. 5.4.

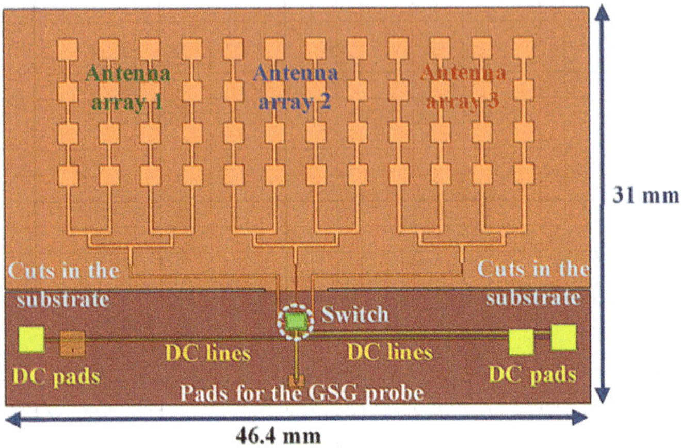


(a)

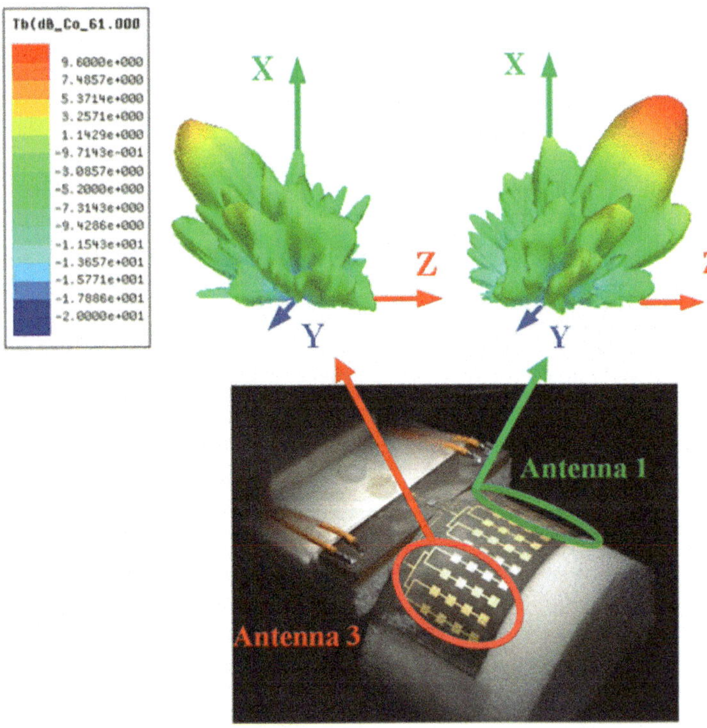


(b)

**Fig. 5.5** **a** Configuration of the beam switching antenna at mm-wave frequency and **b** radiation patterns of antenna 1 and 3 ([6], Reprinted with permission from IEEE)

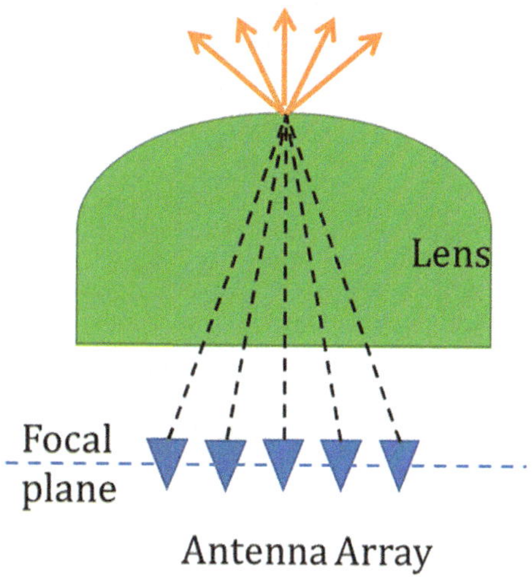


**Fig. 5.6** Integrated lens antenna array

## 5.4 Metamaterial Based Wide-Scan Angle Beam Switching Antenna

Metamaterials can replace the bulky dielectric lens for beam steering applications resulting in very compact antenna system. The metamaterial high refractive index surface presented in Chap. 4, Sect. 4.3.2 for beam tilting applications can be utilized with an array of end-fire antennas to achieve wide-scan radiation coverage. In addition to the deflection of radiation beam with the offset loading of the metasurface, the gain enhancement can also be achieved.

The effect of metamaterial loading on the end-fire antenna gain is studied in this Chapter. A comparison of the results between the antenna without and with metamaterial CLL unit cell array is carried out. The single element dipole antenna with metamaterial loading covers entire band from 26 to 30 GHz with input reflection coefficient $<-10$ dB as shown in Fig. 5.8. The effect of loading metamaterial on the radiation characteristics is depicted in the Fig. 5.9a which shows the radiated power from the dipole antenna with and without metamaterial loading, the radiated power flow is more concentrated in the loading area which is due to the higher effective refractive index due to printing of metamaterial CLL array. Therefore, by printing the metamaterial unit cell array, the radiated electromagnetic energy from the dipole antenna is focused in the region which results in enhanced gain. This is verified by the phase profile of the radiated electric field $(E_x)$ from the dipole antenna at $y = 6$ mm with and without metamaterial CLL printing as shown in Fig. 5.9b. The phase plot shows the radiated field in case of loading with CLL array, the radiated power would be along y-axis only as the power flows along the normal to the phase profile of the fields, whereas in case of no loading, radiated power is scattered in the xy-plane resulting in low gain. Thus, with the loading of the CLL array, the forward gain of

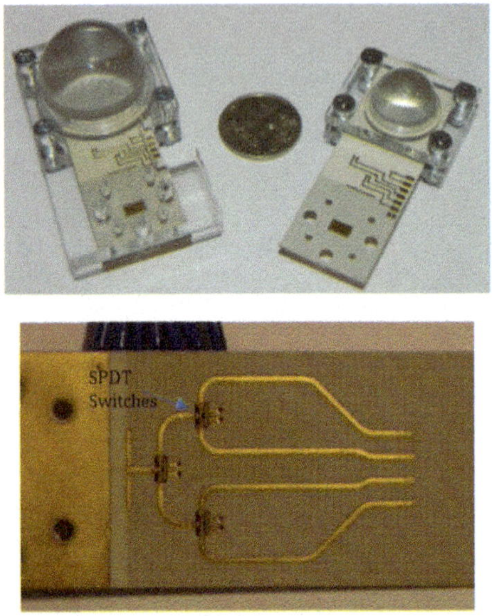


(a)

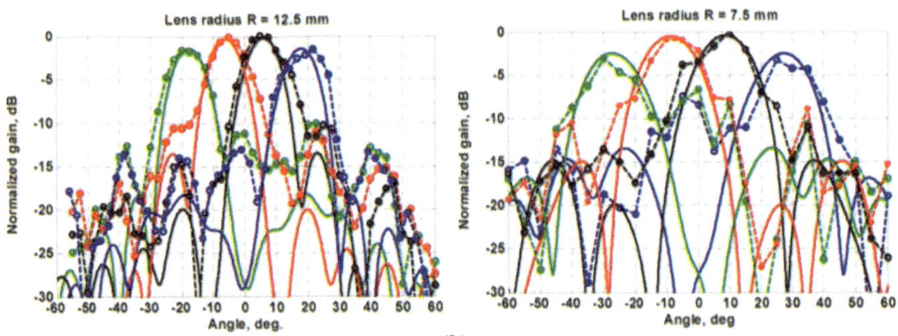


(b)

**Fig. 5.7** **a** Configuration of the electronic beam switching antenna with the lens and **b** radiation patterns with 12 mm and 7.5 mm lens radius ([10], Reprinted with permission from IEEE)

the dipole antenna is enhanced due to quasi-planar wave front of the radiated waves that results due to no phase accumulation of the waves along y-axis.

The measured and simulated E-plane radiation pattern of the dipole antenna with metamaterial loading is shown in Fig. 5.10a along with the radiation pattern of dipole antenna without loading. The printing of metamaterials results in decreased 3-dB beamwidth to 41° from 64.5° which translates to a gain enhancement of about 5.9 dB. The measured and simulated gain of the antenna with and without metamaterials is plotted in Fig. 5.10b as a function of frequency. It is observed that if the metamaterial unit cell array is placed symmetrical about the y-axis, the emitted EM rays from the dipole antenna are launched into the high refractive index region, which results

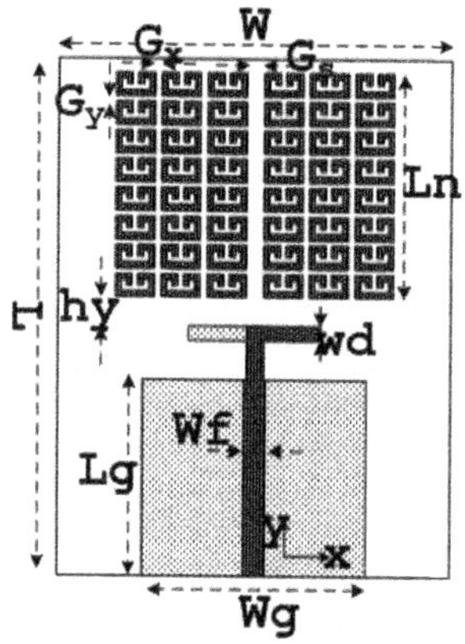


(a)

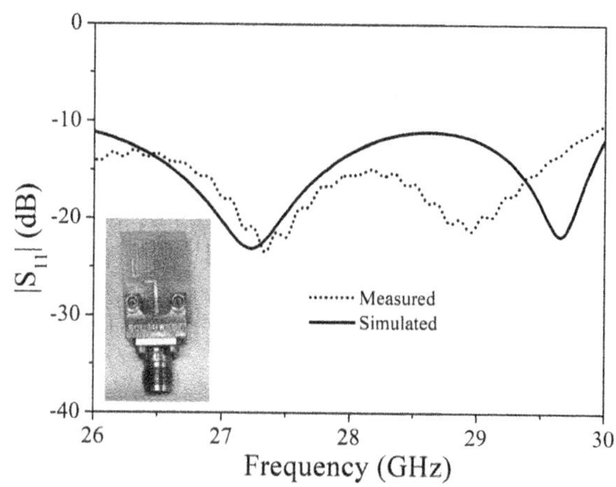


(b)

**Fig. 5.8** Dipole antenna with metamaterial loading **a** antenna geometry (W = 12, L = 18, Wg = 8, Lg = 6.8, Ln = 7.8, wd = 0.5, Wf = 0.77, Gs = 1.24, hy = 1) all dimensions are in millimetres, **b** simulated and measured $|S_{11}|$ of the dipole w/ML ([13], Reprinted with permission from IEEE)

in the focused end-fire radiation with peak gain of about 11 dBi in simulations. The peak gain of about 10 dBi is obtained in measurements, the sudden roll off of the measurement gain curve is attributed to fabrication tolerances of the prototype antenna. It was observed that some of the CLL unit cell were not correctly printed on the substrate and had uneven gap widths which directly affects the $S_{21}$ resonance dip frequency of the CLL unit cell.

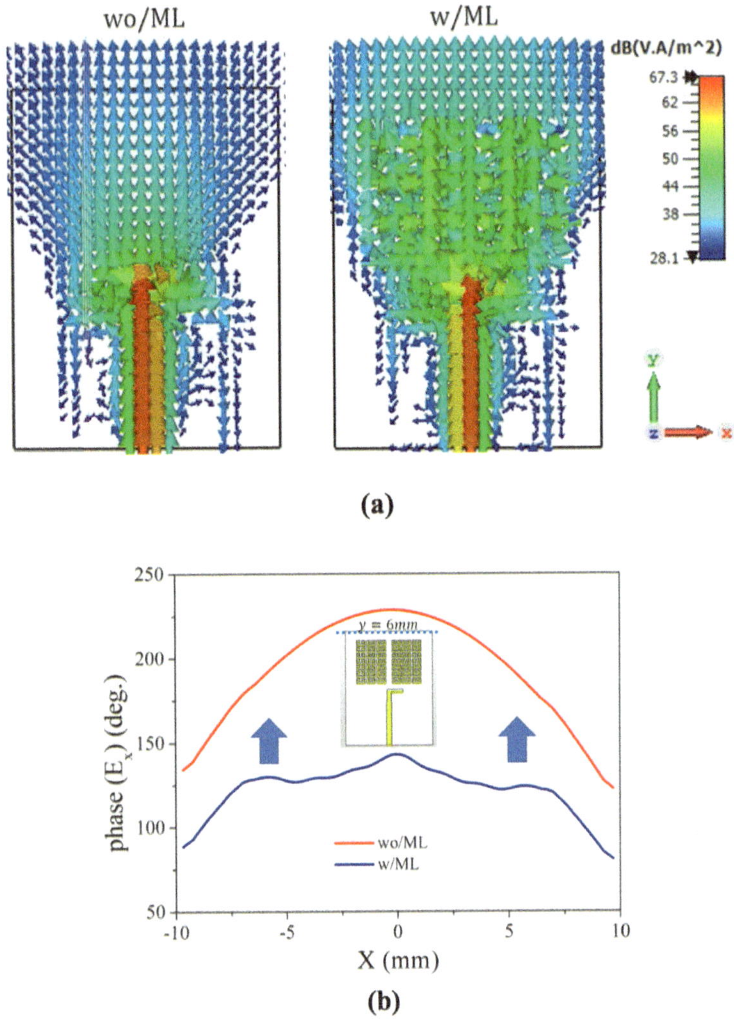


**Fig. 5.9** **a** Power flow distribution of the dipole antenna with (w) and without (wo) metamaterial loading **b** phase profile of electric field w/wo ML

### 5.4.1 *Wide Scan Antenna Array with Metamaterial Loading*

The gain enhancement and beam titling using the integrated CLL unit-cell array in front of the dipole antenna is employed to achieve wide-scan radiation coverage. The quasi Yagi-Uda antenna element is printed on the same Neltec substrate with a copper thickness of 17 μm and loss tangent of 0.0009. The antenna element employs a single director, which is printed on the top side of the substrate and is separated by a spacing of $s_d$ = 1.8 mm from the driving dipole. The quasi Yagi-Uda antenna

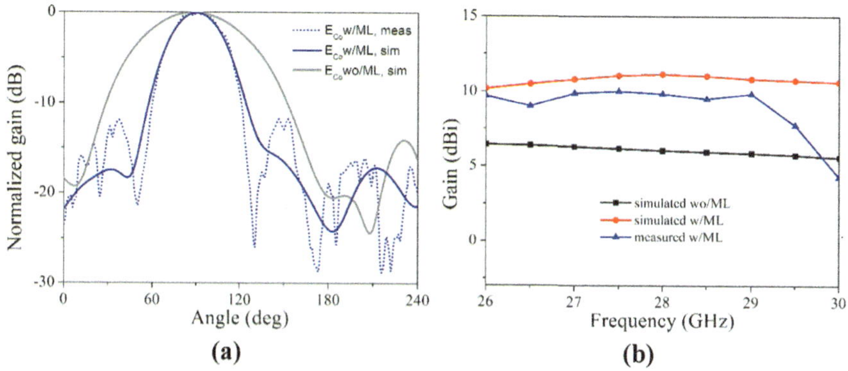


**Fig. 5.10** **a** Measured and simulated E-plane radiation pattern at 28 GHz in xy-plane with (w/ML) and without metamaterial loading (wo/ML), **b** gain curve comparison w/ML and wo/ML ([13], Reprinted with permission from IEEE)

element is designed initially using the dimensions from the table for maximum gain in air [14], and then scaling them to compensate for effective dielectric constant of the substrate. A microstrip feed of width, $w_f = 0.77$ mm is used to realize $Z_o$ of 50 Ω for excitation.

The three-element quasi Yagi-Uda antenna is printed on the substrate with a spacing of 5.4 mm ($0.5\lambda_o$ at 28 GHz) between the centres of the adjacent antennas A1 and A2 (and A2 and A3) as shown in Fig. 5.11. The antenna elements A1 and A3 are tilted by an angle, $\alpha = \mathbf{15}°$ towards the metamaterial region as shown in the figure, this tilting of the driving dipole antennas launches the maximum radiated EM rays into the metamaterial region to tilt them further and increase the gain of antennas A1 and A3. The effect of tilting the antenna element A1 by an angle $\boldsymbol{\alpha}$ on different radiation parameters is summarized in Table 5.1. It can be seen from the table that by tilting the antenna, the radiation direction is also tilted which results in the wide scan angle radiation coverage. Also, the gain and side-lobe level (SLL) are improved when the antenna is tilted.

The proposed metamaterial unit-cell array discussed in Chap. 4, Sect. 4.3.1 is loaded to the array of three quasi-Yagi-Uda antennas to increase the scan angle and gain of the individual antenna elements. The metamaterial region produces lensing effect and focuses the E-field from antenna elements. The metamaterial lens is printed on the same substrate on which three antenna elements are printed, making the complete system planar and compact. The EM waves with x-polarization enters the metamaterial region in the +y direction owing to the antenna configuration and geometry, with E-field and H-field of the wave along x- and z-directions respectively. The loading of metamaterial unit cell layers onto the antenna does not affect the reflection coefficient much, when these layers are printed at a distance of $h_y = 1.2$ mm from the antenna director.

The effect of loading metamaterial on reflection coefficient of the antenna is studied and the antenna is optimized for a gap 'Gp' between two CLL arrays which

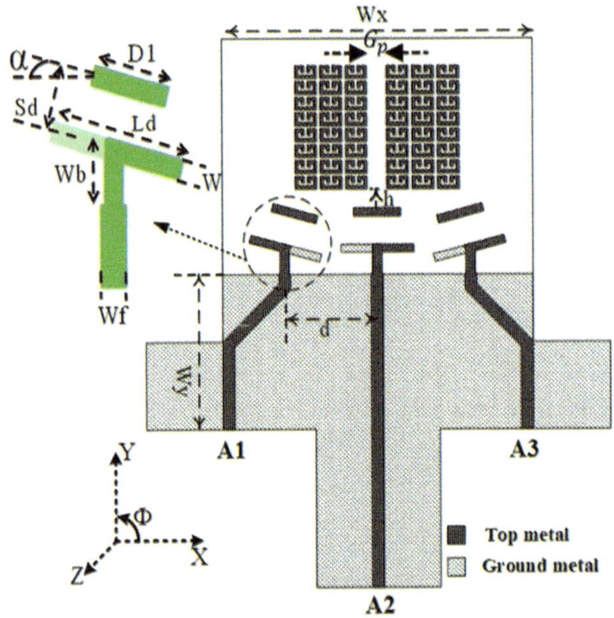


**Fig. 5.11** Three element antenna with metamaterial loading (dimensions are in mm): $Wx$ = 20, $Wy$ = 9.8, $W$ = 0.5, $Wb$ = 1.9, $Ld$ = 4.7, $D1$ = 3, $h$ = 1.2, $d$ = 5.4, $Ld$ = 4.7, $Gp$ = 1.24) ([15], Reprinted with permission from IEEE)

**Table 5.1** The effect of tilting antenna-A1

| α | Gain (dBi) | SLL (dB) | Beam tilt in main lobe (90 − Φ) | FBR |
|---|---|---|---|---|
| 0° | 10.6 | −9.5 | 27° | 19 |
| 5° | 10.66 | −9.7 | 26° | 19.4 |
| 10° | 10.64 | −9.9 | 27° | 19.7 |
| 15° | 10.9 | −10.7 | 29° | 19.8 |
| 20° | 10.4 | −10 | 28° | 15.8 |

controls the reflection coefficient. The effect of gap on $|S_{22}|$ of the antenna-A2 is plotted in Fig. 5.12. It can be seen from the figure that by loading, the $|S_{22}|$ gets affected but by choosing a gap, Gp = 1.24 mm between the CLL arrays, antenna-A2 covers the entire 25 to 30 GHz band with $|S_{22}| < -10$ dB. Also, the effect of loading the metamaterial unit cell layers on different antenna parameters are summarized in Table 5.2. The E-field distribution at 28 GHz is also studied using CST MWS to get an insight into the lensing property of the metamaterial layers. The E-field distribution can be seen from the snapshot in Fig. 5.13 when excited at different ports in the xy-plane. From this figure, it can be observed that the isolation between antennas is slightly affected by loading, but it results in enhanced gain. The E-field distribution in the metamaterial region is symmetrical about y-axis when antennas

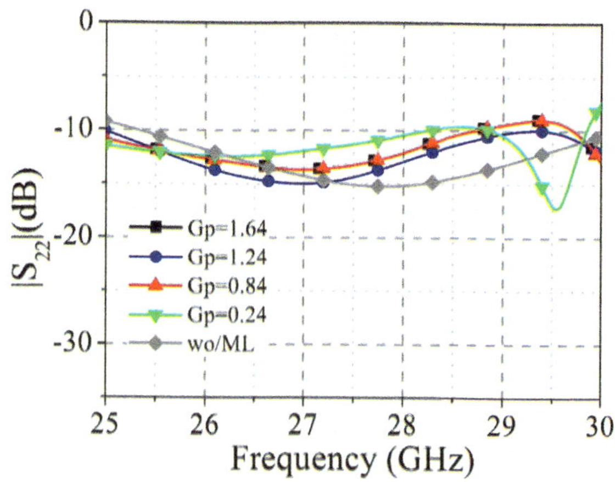


**Fig. 5.12** Effect of gap 'Gp' on $|S_{22}|$ of the antenna-A2 ('Gp' in millimetres) ([15], Reprinted with permission from IEEE)

**Table 5.2** The effect of metamaterial loading

| | Without ML | With ML |
|---|---|---|
| Gain (dBi) | A1 = A2 = A3 = 6.5 | A1 = A3 = 10.85A2 = 11.31 |
| Beam scan angle | 47° | 64° |
| Front-to-back ratio (FBR) | 14 | 18 |
| Side-lobe level (SLL) | −16.3 dB | −10.6 dB |
| Envelop correlation coefficient (ECC) | 2E−3 | 1E−2 |
| Total efficiency | >86% | >80% |

A1 and A3 are excited and the E-field distribution when A2 is excited is more focused along the end-fire direction, which results in the gain of the A2 element being more than A1 and A3. As seen from Fig. 5.13, the E-field radiated from antenna-A1 is tilted to right by high refractive index region along, $\Phi = \mathbf{60°}$ and similarly E-field from A3 is tilted along, $\Phi = \mathbf{120°}$. The E-field from the A2 is entering the region along the normal and is radiated along the same direction. The focusing effect of the E-field can also be seen from the figure when the antenna array is loaded with the metamaterial unit-cell array. Figure 5.14a shows the radiation pattern (E-plane) of the antenna with and without metamaterial loading at 28 GHz for all the three ports. It is clear from the figure that the HPBW reaches **36.1**° from **61.8**° for antenna-A1, **32.7**° from **76**° for antenna-A2 and **35.6**° from **67.6**° for antenna-A3 when loaded with metamaterial array. The 3D radiation pattern of the wide scan antenna array for the three ports is plotted in Fig. 5.14b at 28 GHz. As observed, the radiation is focused along three different directions for the ports with improved gain along the direction.

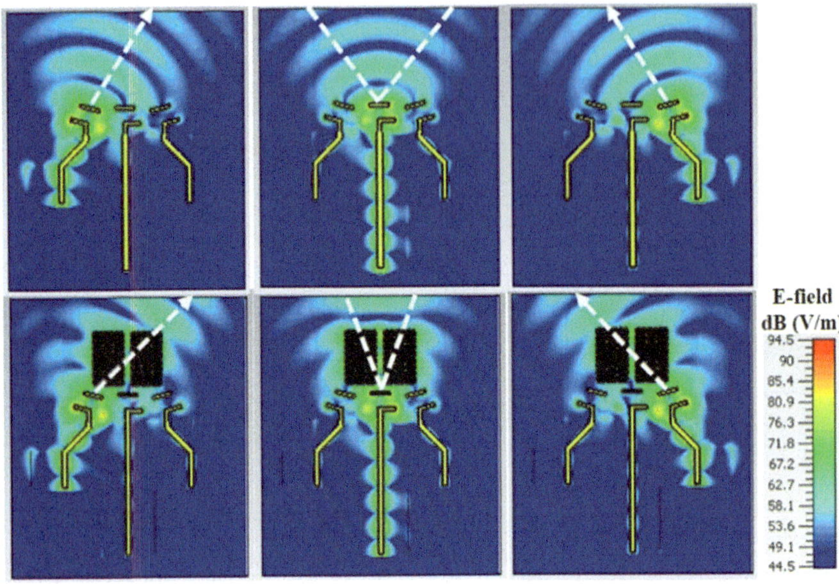


**Fig. 5.13** E-field distribution at 28 GHz of three-element antenna array without and with metamaterial loading when three different ports are excited ([15], Reprinted with permission from IEEE)

Thus, the net effect of printing metamaterial unit-cell array on the substrate is: (i) an increase in the scan angle by bending the EM rays from antenna elements A1 and A3 in the direction **60**° and **120**° and (ii) an increase in the gain by concentrating the E-field from antennas into specific directions. The proposed antenna with wide scan can be useful in spatial beamforming MIMO applications to increase the throughput of the system when all three antennas are used independently. The proposed antenna can also be used in beam switching applications when employed with a SP3T switch to scan the radiation in the E-plane. The beam switching capability here is studied manually by exciting individual ports at a time.

### *5.4.2 Prototype and Measurement Results*

The prototype three-element antenna with proposed metamaterial loading is fabricated using the normal photolithography process in our laboratory. The proposed wide-scan angle three-element antenna array is fabricated (Photograph shown in Fig. 5.15a) and fed using 2.92 mm end launch connectors. The prototype is tested using Agilent E8364C PNA Microwave Network Analyzer and predicted results are verified. The measured reflection coefficient is compared with the simulated results in Fig. 5.15b. It is seen that all the three elements loaded with metamaterial layer

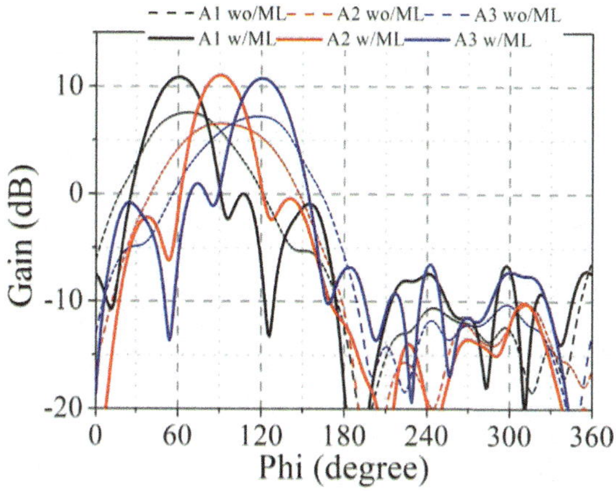


(a)

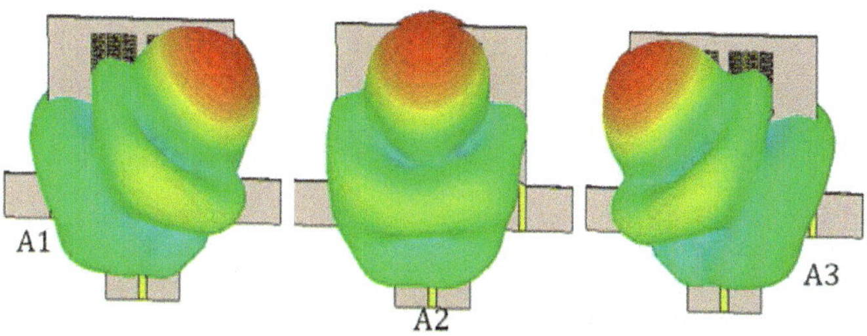


(b)

**Fig. 5.14** **a** E-plane radiation pattern of the 3-port antenna with (w/ML) and without metamaterial loading (wo/ML) at 28 GHz, **b** 3D radiation pattern for antennas-A1, A2 and A3 ([15], Reprinted with permission from IEEE)

covers the frequency range 25–30 GHz with $|S_{11}| < -10$ dB. Also the mutual coupling between the three antennas is presented in Fig. 5.15c, as evident from the figure, a low mutual coupling of −16 dB between the antennas in the frequency range 25–30 GHz is observed. Low mutual coupling value obtained is very useful for MIMO applications, resulting in improved capacity of the overall communication system.

The measured and simulated gain of the individual antenna elements A1 and A2 with and without metamaterial loading is shown in Fig. 5.16. The gain of the individual antenna elements A1, A2 and A3 without metamaterial loading is around 6 dBi. The effect of incorporating metamaterial layer onto the substrate increases the gain of individual antenna elements by around 5 dB, and the gain is constant over the frequency range 25–30 GHz. The measured gain of A1 (same as A3) is around 10.85 dBi and that of A2 is 11.3 dBi at 28 GHz. The gain of the antenna is stable and a 3-dB gain bandwidth of about 18% is achieved by metamaterial loading.

The simulated and measured radiation pattern of the proposed antenna with metamaterial loading at 27, 28 and 29 GHz is shown in Fig. 5.17. The measurements were

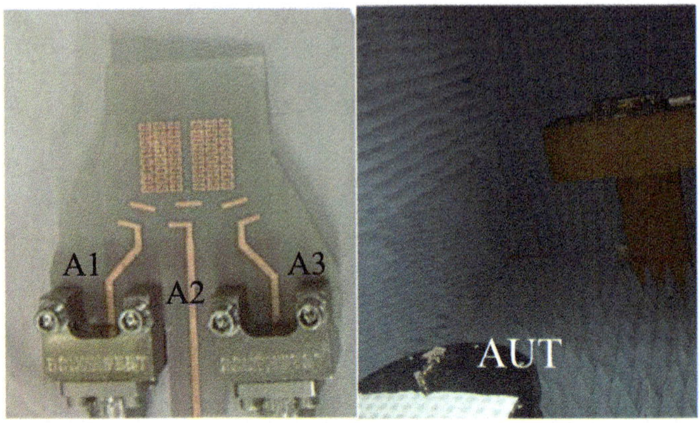


**(a)**

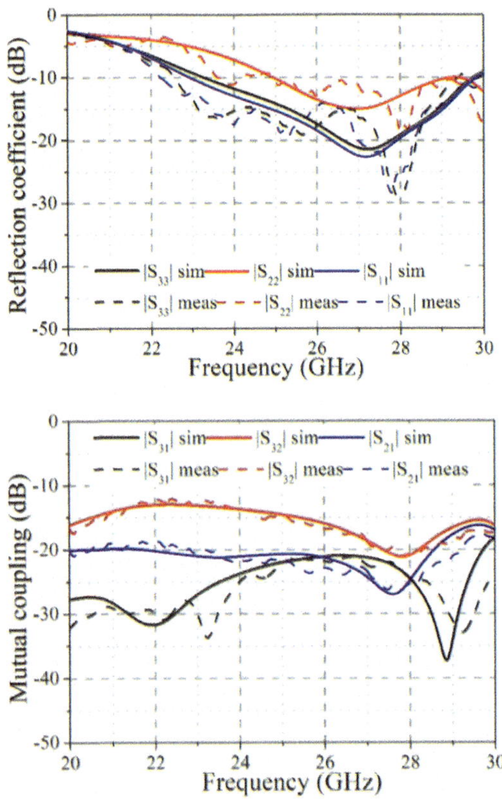


**(c)**

**Fig. 5.15** **a** Photograph of the fabricated prototype and measurement setup, **b** measured and simulated reflection coefficient **c** measured and simulated coupling between all the ports ([15], Reprinted with permission from IEEE)

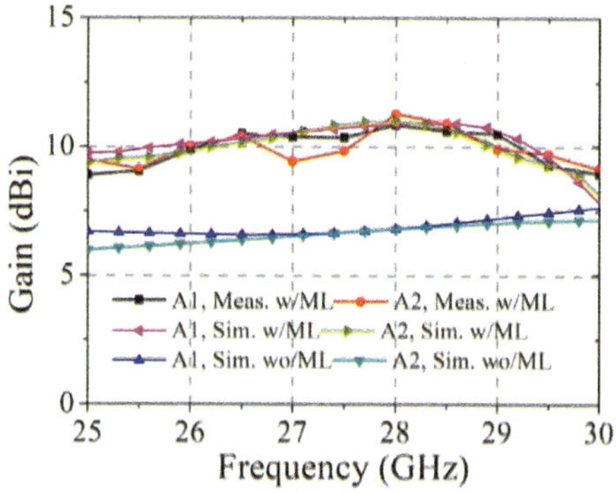


**Fig. 5.16** Comparison of gain of the antenna element A1 and A2 with (w/ML) and without (wo/ML) metamaterial loading ([15], Reprinted with permission from IEEE)

carried out in a far-field type anechoic chamber. The radiation patterns in xy-plane show pattern diversity as the pattern of antennas are directed at different angles, $\mathbf{\Phi} = \mathbf{58}^\circ$, $\mathbf{90}^\circ$ and $\mathbf{122}^\circ$, covering an angle of $\mathbf{64}^\circ$. The incorporation of metamaterial layers has increased the beam scan angle to $\mathbf{64}^\circ$ with non-overlapping 3-dB beamwidth for the three ports as can be seen from Fig. 5.17.

This is useful for applications where each antenna serves a particular angular space which was not possible with the overlapping beamwidths of the three-port antenna without the metamaterial layers. This wide scan angle of $\mathbf{64}^\circ$ is useful in applications where the transmitter and receiver location are not known. The wide scan antenna array performance characteristics are compared with existing state of the art beam steering antennas and the comparison is presented in Table 5.3. The proposed antenna has wider 3-dB gain bandwidth with peak gain of 11 dBi at a compact size and covering 58° to 122° scan angle in azimuthal plane.

## 5.5 Conclusion

This Chapter presents beam switching antenna arrays for mmWave wireless communication systems. The conventional beam switching technique using Butler matrix and integrated lens antennas are discussed in detail. It has been observed that the losses in the feed network and the cost of phase shifters at mmWave frequencies restrict the use of Butler matrix at these frequencies. Later in the Chapter, integrated lens antennas (ILA) with dielectric lens for beam steering applications are introduced. The ILA provides highly directional radiation beams; however, the steering range is limited. The implementation of metamaterial high refractive impedance surface with the end-fire antenna array is finally presented for wide-scan beam switching applications replacing the bulky dielectric lens in ILA.

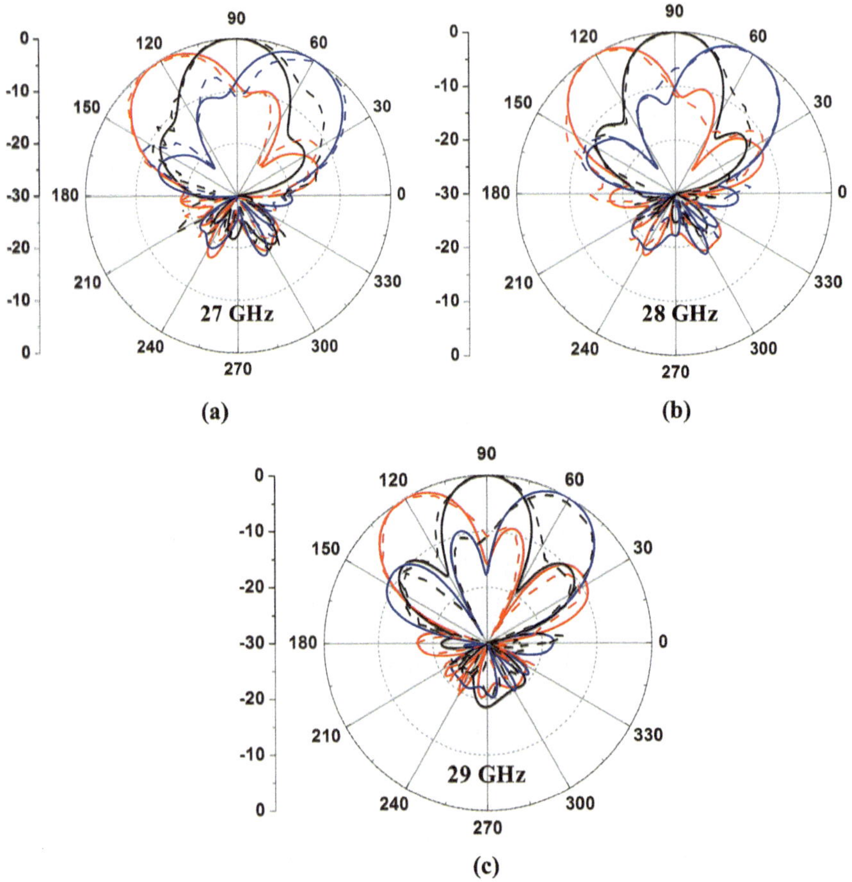


**Fig. 5.17** Measured and predicted normalized gain pattern in E-plane of three-element antenna array with metamaterial loading at **a** 27 GHz, **b** 28 GHz and **c** 29 GHz (Blue: Port1, Black: Port2 and Red: Port3; solid line: simulated, dashed line: measured) ([15], Reprinted with permission from IEEE)

**Table 5.3** Comparison of the reported literature

| Refs. | Freq. band (GHz) | Peak gain (dBi) | 3-dB gain bandwidth | E-plane scanning | Volume | Type |
|---|---|---|---|---|---|---|
| [16] | 57–64 | 12.4 | 11 | −35° to 35° | $1.09\lambda_o^3$ at 60 GHz | 2-port E and H-plane ML antenna |
| [11] | 50–70 | 20 | 33 | 60° to 120° | $131\lambda_o^3$ at 60 GHz | 3-port dielectric loaded antenna |
| [17] | 57–64 | 10 | 11 | −26° to 26° | $0.04\lambda_o^3$ at 60 GHz | 2-port ML antenna |
| [18] | 33–37.8 | 6 | 12 | No | $0.11\lambda_o^3$ at 35 GHz | SIW cavity antenna |
| [19] | 25–37.5 | 10 | 17 | No | $0.4\lambda_o^3$ at 28 GHz | 4-port defected ground T-antenna |
| Wide-scan metamaterial antenna [15] | 25–30 | 11.3 | 18 | 58° to 122° | $0.05\lambda_o^3$ at 28 GHz | 3-port ML antenna |

# References


1. Neekzad B, Sayrafian-Pour K, Baras JS (2008) Clustering characteristics of millimeter wave indoor channels. In: 2008 IEEE wireless communications and networking conference, pp 1217–1222. http://doi.org/10.1109/WCNC.2008.219
2. Chryssomallis M (2000) Smart antennas. IEEE Antennas Propag Mag 42(3):129–136. https://doi.org/10.1109/74.848965
3. Fakoukakis FE, Kaifas TN, Vafiadis EE, Kyriacou GA (2015) Design and implementation of Butler matrix-based beam-forming networks for low sidelobe level electronically scanned arrays. Int J Microw Wirel Technol 7(1):69–79. https://doi.org/10.1017/S1759078714000403
4. Mosca S, Bilotti F, Toscano A, Vegni L (2002) A novel design method for Blass matrix beam-forming networks. IEEE Trans Antennas Propag 50(2):225–232. https://doi.org/10.1109/8.997999
5. Wang H, Zhang Z, Feng Z (2012) A beam-switching antenna array with shaped radiation patterns. IEEE Antennas Wirel Propag Lett 11:818–821. https://doi.org/10.1109/LAWP.2012.2206362
6. Semkin V et al (2016) Beam switching conformal antenna array for mm-wave communications. IEEE Antennas Wirel Propag Lett 15:28–31. https://doi.org/10.1109/LAWP.2015.2426510
7. Buttgenbach TH (1993) An improved solution for integrated array optics in quasi-optical mm and submm receivers: the hybrid antenna. IEEE Trans Microw Theory Tech 41(10):1750–1760. https://doi.org/10.1109/22.247920
8. Räisänen AV et al (2013) Studies on E-band antennas and propagation. In: 2013 Loughborough antennas propagation conference (LAPC), pp 176–180. http://doi.org/10.1109/LAPC.2013.6711876
9. Artemenko A, Maltsev A, Maslennikov R, Sevastyanov A, Ssorin V (2011) Beam steerable quartz integrated lens antenna for 60 GHz frequency band. In: Proceedings of the 5th European conference on antennas and propagation (EUCAP), pp 758–762

10. Artemenko A, Maltsev A, Mozharovskiy A, Sevastyanov A, Ssorin V, Maslennikov R (2013) Millimeter-wave electronically steerable integrated lens antennas for WLAN/WPAN applications. IEEE Trans Antennas Propag 61(4):1665–1671. https://doi.org/10.1109/TAP.2012.2232266
11. Briqech Z, Sebak AR, Denidni TA (2016) Wide-scan MSC-AFTSA array-fed grooved spherical lens antenna for millimeter-wave MIMO applications. IEEE Trans Antennas Propag 64(7):2971–2980. https://doi.org/10.1109/TAP.2016.2565704
12. Filipovic DF, Gauthier GP, Raman S, Rebeiz GM (1997) Off-axis properties of silicon and quartz dielectric lens antennas. IEEE Trans Antennas Propag 45(5):760–766. https://doi.org/10.1109/8.575618
13. Wani Z, Abegaonkar MP, Koul SK (2017) Gain enhancement of millimeter wave antenna with metamaterial loading. In: 2017 international symposium on antennas and propagation (ISAP), pp 1–2. http://doi.org/10.1109/ISANP.2017.8229014
14. Balanis CA (2009) Antenna theory: analysis and design, 3rd edn. Wiley India Pvt. Limited [Online]. Available: https://books.google.co.in/books?id=2SJ0CgAAQBAJ
15. Wani Z, Abegaonkar MP, Koul SK (2019) Millimeter-wave antenna with wide-scan angle radiation characteristics for MIMO applications. Int J RF Microw Comput Aided Eng 29. https://doi.org/10.1002/mmce.21564
16. Dadgarpour A, Zarghooni B, Virdee BS, Denidni TA (2016) One- and two-dimensional beam-switching antenna for millimeter-wave MIMO applications. IEEE Trans Antennas Propag 64(2):564–573. https://doi.org/10.1109/TAP.2015.2508478
17. Dadgarpour A, Zarghooni B, Virdee BS, Denidni TA (2015) Beam-deflection using gradient refractive-index media for 60-GHz end-fire antenna. IEEE Trans Antennas Propag 63(8):3768–3774. https://doi.org/10.1109/TAP.2015.2438396
18. Zhang Y, Chen ZN, Qing X, Hong W (2011) Wideband millimeter-wave substrate integrated waveguide slotted narrow-wall fed cavity antennas. IEEE Trans Antennas Propag 59(5):1488–1496. https://doi.org/10.1109/TAP.2011.2123055
19. Jilani SF, Alomainy A (2018) Millimetre-wave T-shaped MIMO antenna with defected ground structures for 5G cellular networks. Antennas Propag. IET Microw 12(5):672–677. https://doi.org/10.1049/iet-map.2017.0467

# Chapter 6
# Millimeter Wave MIMO Antennas

## 6.1 Introduction

The multiple-input-multiple-output (MIMO) system has multiple transmitting and receiving antennas to improve the performance of the communication systems. These systems significantly improve the wireless link range and data throughput. Since the mmWave wireless communication has a limited range and reliability, therefore implementation of MIMO at mmWaves would result in improved reception and spectral efficiency. As we have discussed in previous chapters that the mmWave wireless communication faces issues like higher path loss and beam alignment. The solution to these issues has been addressed in Chaps. 3–5.

This chapter focusses on the issue of shadowing of mmWave signals by humas or any other object which results in loss of communication link. The penetration loss of the mmWave 28 GHz signal for tinted glass and bricks is of the order of 40 dB and 28 dB respectively [1]. The huge penetration loss makes it difficult to maintain the communication link with sufficient SNR. However, for non-LOS communication, the signal power loss can be reduced by employing multiple antenna system having multiple radiation beams at the receiver side by combining received power which eventually improves the SNR. The design and implementation of MIMO antennas at mmWave frequencies greatly benefit from the smaller antenna size which helps in integrating large number of antennas in a small footprint.

## 6.2 mmWave MIMO Antennas

Typical antennas that are used for mmWave communications are reflector, lens and horn antennas [2]. These types of antennas provide high gain, however they are not preferred for commercial handheld applications because of their large size and cost. The most simple and easy to fabricate using photolithography process are printed

 
S. K. Koul and Z. Wani, *Novel Millimetre Wave Antennas for MIMO and 5G Applications*, Lecture Notes in Electrical Engineering 819,
https://doi.org/10.1007/978-981-16-7278-1_6

circuit antennas. These antennas which include microstrip patch or dipole arrays are low profile, light weight and can be used in small handheld planar devices. As a result, they have attracted attention of many researchers to work on the design of these printed antenna arrays for mmWave communication. In [3], microstrip fed Yagi-Uda antenna is presented for millimetre wave applications. Authors have reported a gain of 9 dB using five directors on the substrate placed in front of the dipole antenna. Extensive research has already been reported on the design of MIMO antennas for wireless 3G and LTE 4G systems. Taking the advantage of MIMO technology to increase the throughput of the next generation mmWave communication, antennas in [4–11] have been proposed for 28 GHz, 38 GHz and 60 GHz bands. In [12], authors have used 2 × 2 spatially multiplexed MIMO antenna configuration to achieve 51 Gb/s of data rate using orthogonal frequency division multiplexing system. Besides the conventional approach to increase the performance of the system for example using antenna arrays in MIMO configuration, there is another promising line of research in which to reduce the signal processing complexity, advanced antenna design approach such as lens antenna array are utilized [13, 14]. Table 6.1 presents the mmWave MIMO antennas reported in the literature for 28 GHz band with their characteristics in terms of impedance bandwidth, isolation, and gain.

**Table 6.1** Antennas for mmWave applications

| References | Type | BW | Isolation | Gain | Antenna size | Feature |
|---|---|---|---|---|---|---|
| [3] | Yagi-Uda antenna | 22–26 GHz | 20 dB | 9 dBi | 40 × 30 $mm^2$ | Planar and efficient, beam switching possible |
| [4] | DRA array | 29.4–31 GHz | 15 dB | 10 dBi | 72 × 39 $mm^2$ | 4 port MIMO |
| [5] | Series fed antenna array | (28/38 GHz) | 30 dB | – | 12 × 13 × 20 $mm^3$ | 12 ports, Covers 360, bulky and not planar |
| [7] | PIFA antenna | 1.55 GHz (28 GHz) | – | 4.5 dBi | 4.2 × 4.4 $mm^2$ | Efficiency 94% |
| [8] | Dipole array | 26–35 GHz | – | 7.5 dBi | – | Beam-steering capable antenna array |
| [15] | Phased array antenna | 26–30 GHz | – | 10.9 dBi | – | Compact phased antenna array for cellular apps |
| [16] | Open ended SIW | 26–30 GHz | – | – | – | 180° and 90° beam-width in E-and H-plane |
| [17] | Dipole antenna excited by slot | 27–36 GHz | – | 10 dBi | – | 1 × 4 antenna array |

Different MIMO antennas have been proposed for mmWave spectrum, most of them use antenna arrays to achieve high gain. In [9], 1-D and 2-D beam steering has been achieved up to 35° degrees in both E- and H-planes and the authors have managed to design dual element MIMO antenna. Also, between the two dipole antennas, magnitude of $S_{12}$ indicates mutual coupling between the two dipole antennas, less than -20 dB over 55–67 GHz has been achieved. The designed antenna covers 60 GHz band and beamforming has been achieved using metamaterial lens placed in front of the antenna. Two dipole antennas have been placed close to each other to give rise to space diversity. Authors have reported gain of 12.4 dBi at 62 GHz at a tilt of 35°, but the efficiency achieved is only 88% using dipole antenna, which normally has an efficiency of over 95%. The antenna has a side lobe level (SLL) of − 10 dB and front-to-back ratio (FBR) of 25 dB.

Three-element slot antenna array with spherical dielectric lens to increase the beam scan and gain enhancement for MIMO application at 60 GHz as shown in Fig. 6.1 is proposed in [11]. This antenna is large and occupies good amount of space (131 $\lambda_o^3$, $\lambda_o$ is the free space wavelength at 60 GHz). In [12], authors have used 2 × 2 spatially multiplexed MIMO antenna configuration to achieve 51 Gb/s of data rate using orthogonal frequency division multiplexing system. Besides the conventional approaches to increase the performance of the system for example using antenna arrays in MIMO configuration, there is another promising line of research in which to reduce the signal processing complexity advanced antenna design approach such as lens antenna array are utilized [13, 14]. The EM lens focusses the incident waves on an antenna depending upon its position. With this arrangement spatial beamforming is achieved. Also this techniques will reduce RF chain cost and the losses associated with phase shifters [14]. In [18], dipole antenna with beam deflection using gradient refractive index metamaterial is reported for 60 GHz applications. This antenna uses individual metamaterial unit cells for each antenna to enhance the gain, however

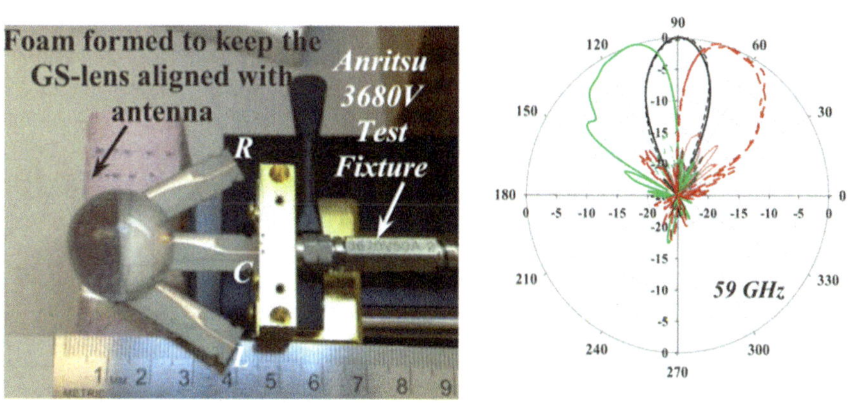


(a) (b)

**Fig. 6.1** **a** Wide-scan MIMO antenna array with spherical grooved-lens and **b** radiation pattern at 59 GHz ([11], Reprinted with permission from IEEE)

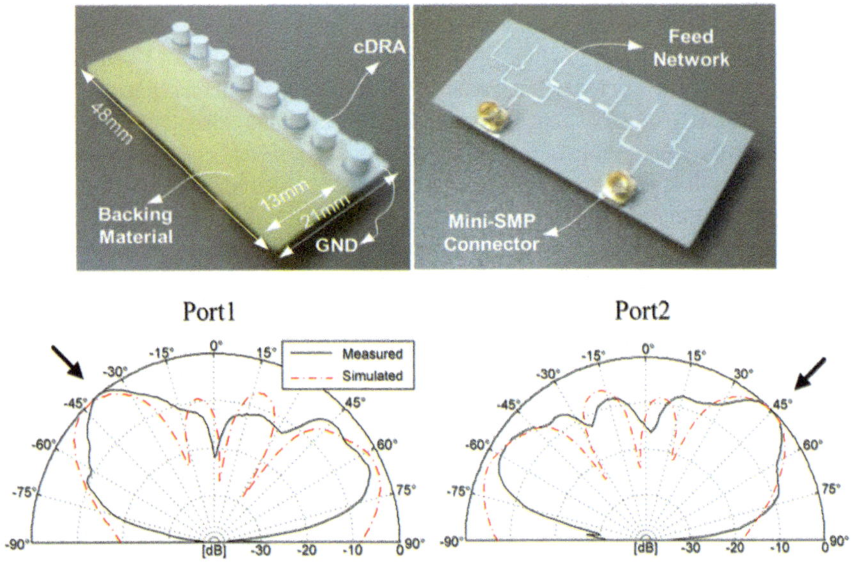


**Fig. 6.2** Two port MIMO antenna for mmWave devices and the radiation pattern at 30 GHz ([19], Reprinted with permission from IEEE)

antenna has limited gain bandwidth of 11%. Also, authors have not mentioned the mutual coupling between the ports and antennas feasibility for MIMO applications. Also in [19], two port MIMO antenna is reported, each antenna consists of four DRA antennas as shown in Fig. 6.2. Despite using array of four DRA antennas on a substrate with area 10.08 $\lambda_0^2$ ($\lambda_o$ = free space wavelength at 30 GHz), authors have achieved a low gain of about 8 dB for mmWave communication.

The mutual coupling between the antennas plays a very important role to achieve good diversity/multiplexing performance in case of MIMO antennas. There are various techniques to improve the isolation between antennas e.g., artificial medias, parasitic elements, and band gap structures. In [20], the mutual coupling between two antipodal slot antennas is reduced using corrugation which results in suppressing the surface waves. The side wall corrugation is done using double negative (DNG) split ring resonators which stop the surface waves going into other port as shown in Fig. 6.3. With the DNG corrugation on side walls of the antennas, 12 dB of improvement in isolation can be observed in Fig. 6.3. This also results in improved efficiency and diversity performance of the antenna.

## 6.3 Compact mmWave MIMO Antenna Design

The design of MIMO antennas for mmWave applications is quite crucial and requires careful consideration. In MIMO antenna system, the mutual coupling between the

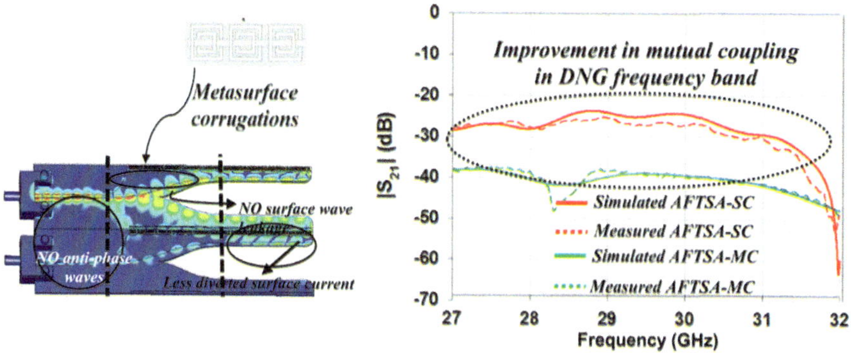


**Fig. 6.3** Metasurface corrugation for reducing mutual coupling between two end-fire antennas at mmWave frequency ([20], Reprinted with permission from IEEE)

antennas plays an important role in determining the performance of the MIMO communication network. Electromagnetic band-gap structure is proposed to reduce the coupling between the antennas in [21]. In [22], single-negative magnetic metamaterials are employed to supress the coupling between two monopole antennas. However, these techniques are not suitable for end-fire antennas. Authors in [23] have enhanced the isolation between two monopole antennas using an array of capacitively loaded loops (CLL) by about 10 dB.

In this chapter, a mmWave MIMO antenna is designed, and each dipole antenna is loaded with high refractive index CLL unit cell array. The loading of CLL unit cell array serves two purposes:

- The CLL array acts a high refractive index region, therefore integration of CLL unit cell array with the dipole antennas enhances the gain of each antenna to about 10 dBi.
- By introducing defects (trimming) in the high refractive index metamaterial unit cell array reduces the mutual coupling between the closely spaced dipole antennas by about 10 dB at 30 GHz.

Low mutual coupling between the antennas and each antenna offering a good gain advocates the feasibility of this antenna for mmWave MIMO applications. The two and four element antenna design for mmWave MIMO applications is discussed below in this section. The design process consists of two main aspects to address their use at mmWaves.

1. Gain enhancement for high signal attenuation
2. Isolation enhancement between the antenna elements, essential for spatial multiplexing

Starting with the two-element dipole antenna configuration with improvement in isolation and then translating antennas to form four element MIMO configuration for better throughput of the communication network is discussed in this section.

### 6.3.1 Two Element MIMO Configuration

The half wave dipole antenna designed has a gain of about 5 dBi at 28 GHz frequency, which is not desirable for communication at mmWave frequency band. High gain antenna design is very crucial, planar Yagi-Uda antennas are most popular end-fire antennas for their high gain [3], here to achieve a gain of around 11 dBi, about 6–7 directors need to be placed which would considerably increase the size of the dipole antenna. Therefore, the gain of the dipole antenna is enhanced by placing the array of CLLs as discussed in Sect. 5.4 of Chap. 5. The CLL-unit cell parameters shown in Fig. 6.4 are extracted from S-Parameters using procedure given in [24] when excited with PEC and PMC boundary conditions in the x- and z-directions. These array of unit cells increases the refractive index of the substrate to about 2.1 (increase in refractive index ~ 42%) and thus focusing of electric field (E-field) is observed. The gain of each dipole antenna is increased to about 10 dBi which is essential for countering propagation losses at 28 GHz frequency band.

The two-element antenna configuration is designed on a Neltec substrate with dielectric constant, $\varepsilon_r = 2.2$, loss tangent $= 0.0009$ and height, $h = 0.254$ mm. The geometry of the proposed antenna with array of metamaterial unit cells is shown in Fig. 6.5. In our design, the metamaterial unit cell array is fed by two dipole antennas in MIMO configuration. The antennas are separated by a distance $S_{dd} = 1.9$ mm ($\lambda_0/5.6$ at 28 GHz) and the mutual coupling between the ports which is due to surface waves, is reduced by adding a stub between the two antennas in the ground plane. This introduces a resonance and the surface currents get accumulated around the stub [26], resulting in enhancement of isolation between the ports. The isolation between the two antennas for different values of the stub length is shown in the Fig. 6.6a. For a stub length of $l_{sb} = 1.8$ mm which is $\lambda/4$ at 28 GHz, the isolation at 28 GHz is increased to 20 dB from 15 dB when a stub was added to the ground.

The other component of the mutual coupling is due to the spatial fields propagating to adjacent antenna through the CLL array. This coupling between the ports is seen to be dependent upon the coupling of the H-field (displacement current) from the high

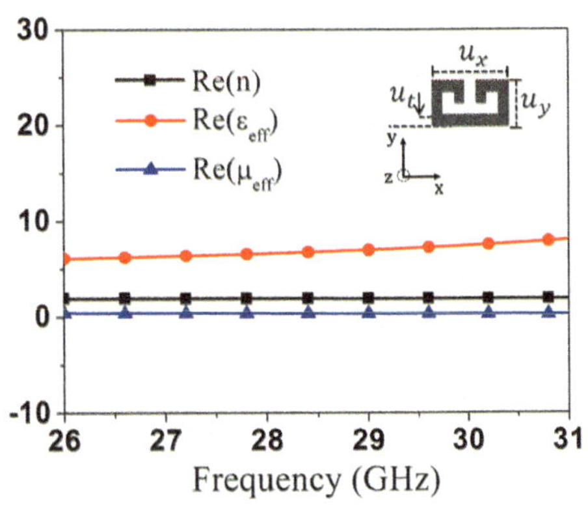


**Fig. 6.4** Extracted real part (Re) of refractive index (n), effective permittivity ($\epsilon_{eff}$) and permeability ($\mu_{eff}$) from S-parameters. ([25], Reproduced courtesy of The Electromagnetics Academy)

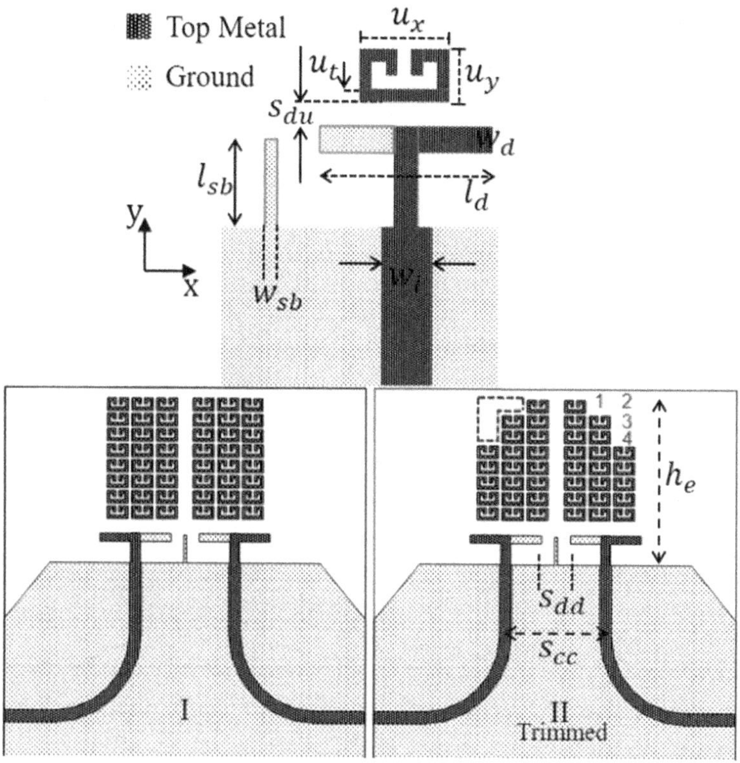


**Fig. 6.5** Geometry of two element MIMO antenna with array of unit cells. Zoomed view of dipole and unit cell along with case-I: without trimming and case-II: with trimming of corners. ($S_{cc}$ = 5.8, $S_{dd}$ = 1.9, $w_i$ = 0.77, $h_e$ = 10.7, $l_d$ = 4.7, $w_d$ = 0.5, $w_{sb}$ = 0.2, $l_{sb}$ = 1.8,$u_t$ = 0.2, $s_{du}$ = 1.2, $u_x$ = 1.4, $u_y$ = 0.8, all dimensions are in millimetres). ([25], Reproduced courtesy of The Electromagnetics Academy)

refractive region on the left (if port 1 is fed) to excite the array on the right, which then propagates back through the CLL array to the adjacent dipole antenna. The effectiveness of trimming of the CLL array is investigated and parametric analysis is done by removing the CLL unit cells, the isolation between the antennas for Case I—without trimming, Case II—with trimming and by removing CLL-1, 2 and 3 is shown in Fig. 6.6b, the graph shows an improvement of 5 dB (at 28 GHz and 10 dB at 30 GHz) in isolation and the overall isolation is greater than 25 dB in the whole frequency band for Case-II. Figure 6.7 shows how the isolation is improved by mitigating the displacement current by trimming the top corners on right and left sides of high refractive index region. The coupling of current can be seen from the CLL array on the right for the Case-I (if port 1 is fed), when no trimming is done.

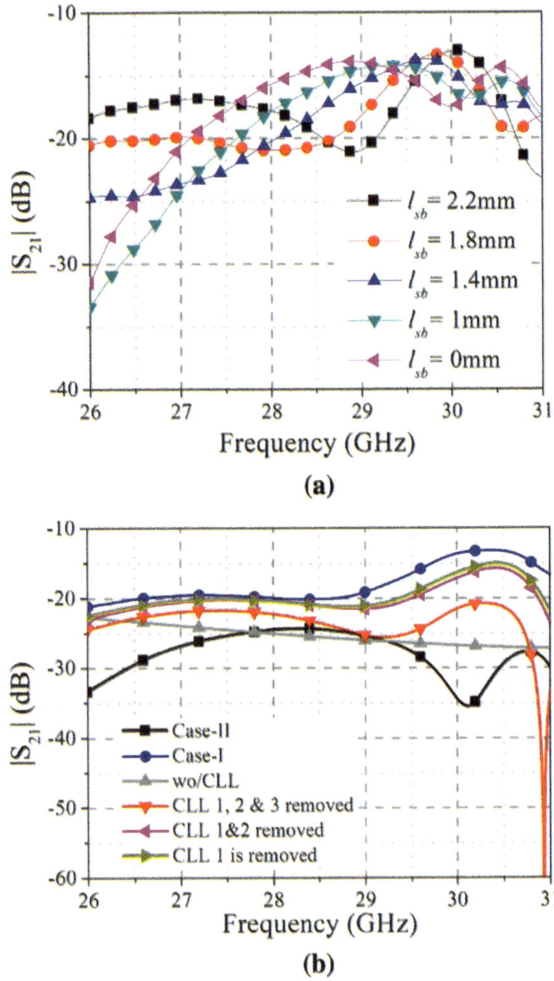


**Fig. 6.6** Isolation between dipole antennas **a** as a function of stub length ($l_{sb}$), **b** without corner trimming (Case I), with corner trimming (Case II), without any CLL array (wo/CLL) and with individual CLL unit cells are removed. ([25], Reproduced courtesy of The Electromagnetics Academy)

### *6.3.2 Two Element MIMO Configuration*

The four port MIMO antenna configuration is shown in Fig. 6.8. All the four antennas are printed on a 10 mil Neltec substrate. The dipole antennas are separated by $S_{cc} = 6$ mm in x-direction and by $S_{hd} = 2.2$ mm along y-direction. All the four antennas and the metamaterial loading to enhance the gain is covering a space of $1.05\lambda_{0@28\text{ GHz}} \times 2.9\lambda_{0@28\text{ GHz}}$ on the substrate without feeds, feed lines are extended to accommodate the end-launch connectors. Each antenna is fed by a 50 Ω microstrip line, and the dipole antenna radiates along the y-direction. The E-field is focused by the array of metamaterial unit cells in different directions depending upon which antenna is fed.

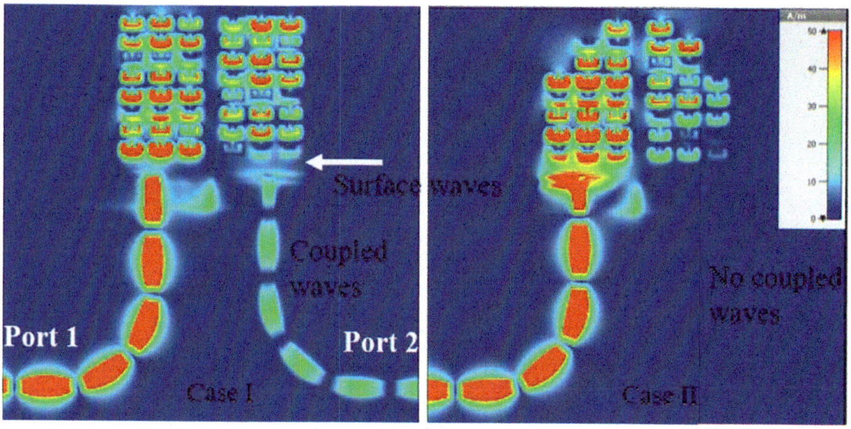


**Fig. 6.7** Snapshot of the H-field in the xy-plane depicting the mutual coupling between the antennas when Port 1 is fed for Case-I and Case-II at 30 GHz. ([25], Reproduced courtesy of The Electromagnetics Academy)

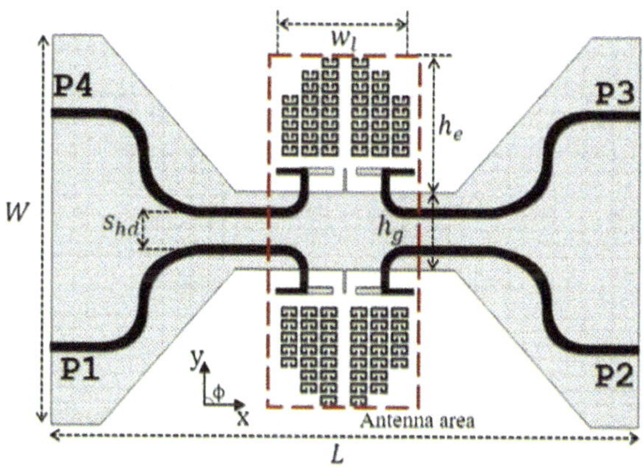


**Fig. 6.8** Proposed four port MIMO antenna configuration geometry. (W = 31, L = 48, $w_l$ = 10.4, $h_e$ = 10.7, $h_g$ = 6.2, $S_{hd}$ = 2.2, all dimensions are in mms) ([25], Reproduced courtesy of The Electromagnetics Academy)

## 6.4 Prototype and MIMO Antenna Performance

The proposed four port antenna for MIMO applications at 28 GHz frequency band is fabricated in-house using the lithography process. The 2.92 mm End Launch Connectors from Southwest Microwave, Inc. are used for measuring the port performance of the prototype antenna shown in Fig. 6.9a using Keysight PNA in our lab. The radiation performance of the antenna is characterized in an anechoic chamber (far-field type) facility available in our Centre and the setup is shown in Fig. 6.9a. The measured and predicted spectral performance of the 4-Port antenna is analysed.

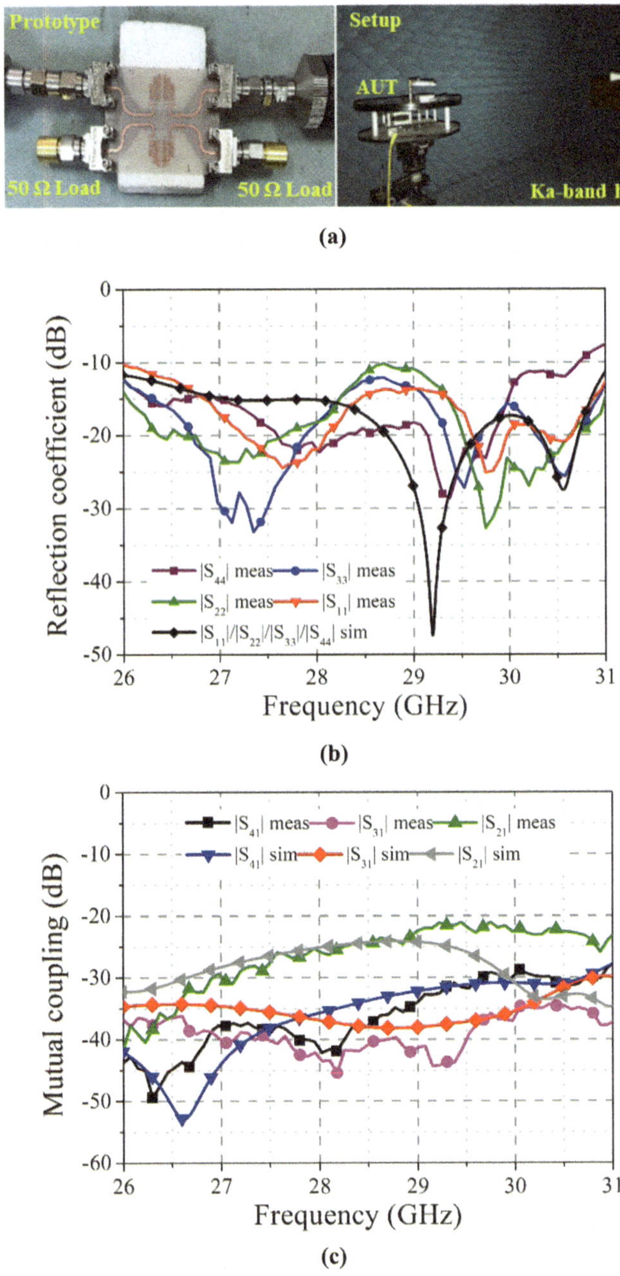


**Fig. 6.9** Port performance of the antenna **a** photograph of the prototype antenna and measurement setup, **b** measured and simulated reflection coefficient and **c** measured and simulated coupling between antennas. ([25], Reproduced courtesy of The Electromagnetics Academy)

The input reflection coefficient for all ports of the antenna in the frequency range 26–31 GHz is shown in Fig. 6.9b and the return loss observed is greater than 10 dB ($|S_{nn}| < -10$ dB). The mutual coupling between the antenna elements is given in Fig. 6.9c. This is very important parameter for the MIMO antennas at 28 GHz, because low coupling (high isolation) between the ports will result in an enhanced efficiency of the individual elements. In addition, it will reduce the power loss due to coupling. Also, the high isolation between antennas makes them independent so that high diversity gains can be achieved. The isolation between the ports is more than 21 dB, which is good enough for MIMO applications at 28 GHz band in such a small space. The radiation pattern of the antenna at 28 GHz when one of the four ports at a time is excited and other ports are terminated in 50 Ω loads in the x–y plane is shown in the Fig. 6.10. The gain patterns show pattern diversity at 28 GHz when excited at different ports. Gain patterns for 29 and 30 GHz are plotted in Fig. 6.11 depicting the stable radiation pattern across the band. As the gain of all antennas is the same, only the measured and simulated gain of the antenna #1 is plotted in Fig. 6.12 over

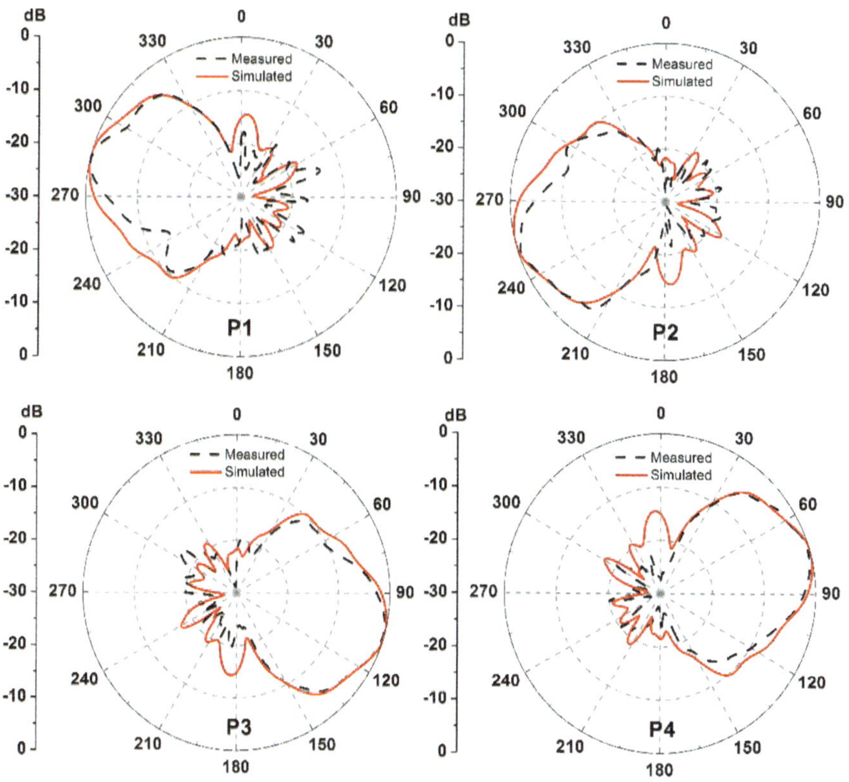


**Fig. 6.10** Measured and simulated normalized gain patterns in xoy-plane at 28 GHz when different ports are excited (Red: Simulated, Black: Measured). ([25], Reproduced courtesy of The Electromagnetics Academy)

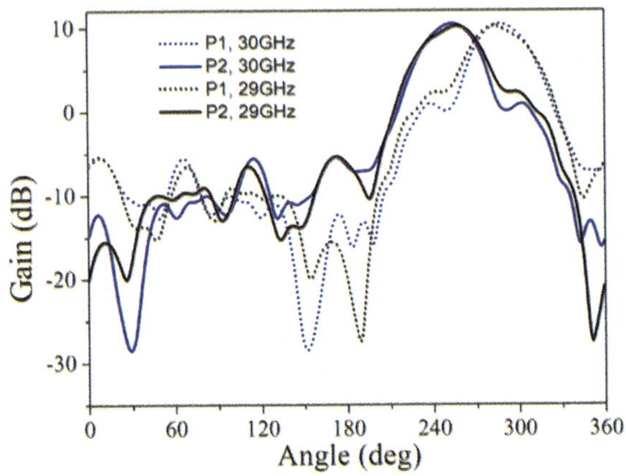


**Fig. 6.11** Gain pattern of the antenna for ports P1 and P2 at 29 GHz and 30 GHz

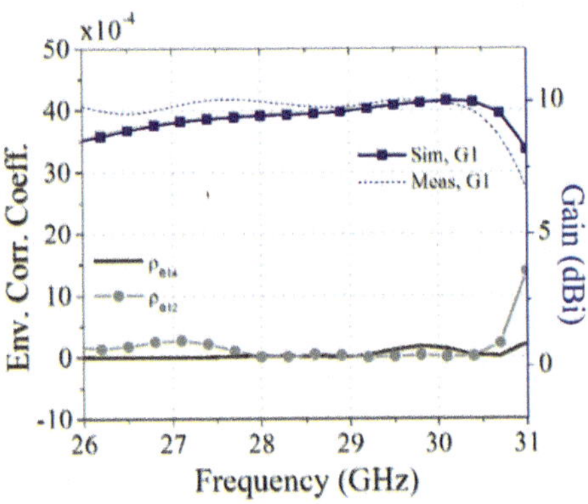


**Fig. 6.12** Envelope correlation coefficient between ports 1 and 2, 1 and 4 and gain of the antenna element 1 versus frequency. ([25], Reproduced courtesy of The Electromagnetics Academy)

the frequency band. It is evident from the plots that the dipole antenna with CLL array is having a stable gain of about 10 dBi in the frequency band.

### 6.4.1 Antenna MIMO Performance

The diversity performance of the antenna for MIMO applications is evaluated in terms of envelope correlation coefficient ($\rho_e$) and is calculated from the 3D radiation patterns ($E_1(\theta, \phi)$ and $E_2(\theta, \phi)$ of the two antennas under isotropic environment consideration using the equation [27]:

$$\rho_e = \frac{\left|\iint [E_1 \cdot E_2^*] d\Omega\right|^2}{\iint |E_1|^2 d\Omega \iint |E_2|^2 d\Omega} \tag{6.1}$$

The diversity performance of the antenna is evaluated using the parameter envelope correlation coefficient (ECC or $\rho_e$), which determines the diversity gain of the multiple antenna systems [28]. The diversity gain will be high for small values of $\rho_e$ which results in the improved system performance. The ECC is plotted between antenna elements 1 and 2 ($\rho_{e12}$) and between antenna elements 1 and 4 ($\rho_{e14}$) in Fig. 6.12. The value of ECC is calculated using far-field results in the CST MWS and is well below than $15 \times 10^{-4}$ in the entire frequency band. The multiplexing efficiency ($\eta_{mux}$) is another parameter to describe the performance of the practical MIMO antenna systems which is the measure of how much more SNR is required to achieve the same capacity as achieved with ideal MIMO antennas (with SNR, $\rho_0$), while using practical MIMO antennas in the same i.i.d channel, and is represented by $\eta_{mux}(\text{dB}) = \rho_0 - \rho_T \leq 0$. Where $\rho_T$ is the SNR required to achieve same capacity in i.i.d channel using practical MIMO antennas and is greater than SNR $\rho_0$ for ideal MIMO antennas.

The multiplexing efficiency matrix ($\eta_{max}$) is used for evaluating the MIMO antennas and is defined as the difference in power required for a MIMO antenna to obtain given capacity relative to ideal MIMO antennas having $\rho_e = 0$ and have 100% total efficiency. For high signal to noise ratio (SNR), the multiplexing efficiency is given by:

$$\eta_{max} = \sqrt{\eta_1 \eta_2 (1 - \rho_e)} \tag{6.2}$$

where $\eta_i$ is the total efficiency of the *i*th antenna [29].

The multiplexing efficiency (ME) is calculated for the designed prototype antennas between ports 1 and 2 for high SNR which is given by (6.2), the ME is plotted in Fig. 6.13a with the total efficiency of the antennas 1 and 2. We can see from the figure that the ME between ports is limited by the total efficiency of the antennas as $\rho_{e12}$ is very small. The minimum ME obtained between the antennas 1 and 2 is − 1.2 dB (75.8%). After calculating the ME, the channel capacity can be calculated and compared with the ideal 2 × 2 MIMO system for different SNR. The prototype antenna has $\rho_{e12} = 0.0015$ and antenna efficiencies, $\eta_1 = \eta_2 = 0.785$, the ergodic channel capacity achieved using the protype MIMO antenna is plotted versus SNR along with the capacity achieved using ideal MIMO antennas system in Fig. 6.13b. It is observed that while using our prototype antenna for 2 × 2 MIMO systems results in a very small capacity loss of 0.76 bps/Hz with respect to ideal MIMO antenna system with ECC = 0 and 100% efficient antennas.

The four port MIMO antenna is characterized, and the performance is compared with the existing literature and is presented in Table 6.2. The edge-to-edge distance between antenna elements is very small ($\lambda_0/6$) with a reasonably good isolation between the ports. To the best of our knowledge, the proposed four port antenna (without extended feed lines) is smaller as compared with most of the reported MIMO antennas at 28 GHz frequency band.

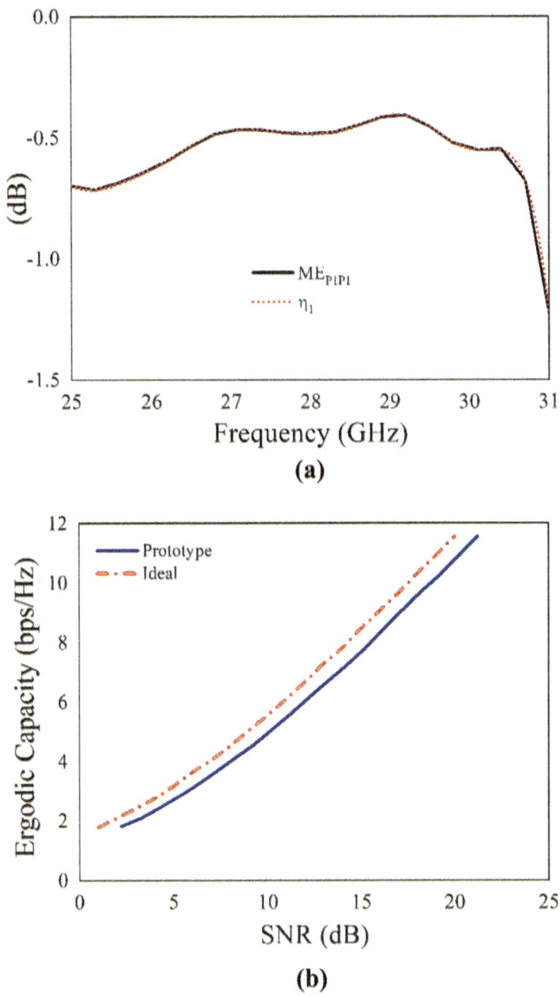


**Fig. 6.13** **a** Multiplexing efficiency between port 1 and 2 and total efficiency, **b** ergodic capacity of the prototype antenna 1 and 2 versus SNR

**Table 6.2** Comparison of the proposed work with other reported works

| Refs. | [19] | [20] | [30] | 4-port MIMO Antenna [25] |
|---|---|---|---|---|
| Total size ($mm^2$) | 48 × 21 | 42 × 85 | NA | 48 × 31 |
| Operating band (GHz) | 29.7–31.5 | 27–32 | 31–40.3 | 26–31 |
| Min Isolation (dB) | 25 | 37.1 | 21 | 21 |
| Peak gain (dBi) | 8.6 | 17.9 | 11 | 10 |
| Edge to edge spacing ($\lambda_0$) | $\lambda_0/3.6$ | N/A | $\lambda_0/1.27$ | $\lambda_0/6$ |
| Number of ports | 2 | 2 | 4 | 4 |

$\lambda_0$ = Free space wavelength at lowest operating frequency

## 6.5 Conclusion

This chapter presents MIMO antennas for mmWave wireless communications. Various MIMO antenna topologies for mmWave applications are analysed in terms of size and diversity/multiplexing performance. Two and four element MIMO antennas at 28 GHz having reasonably good gain and isolation between the ports are presented. It is shown that the isolation between the ports is enhanced by introducing defects in CLL array and ground stub. In the last, the diversity and multiplexing performance is evaluated. The ergodic capacity graph when using the antenna for wireless communication have shown the applicability of antenna for mmWave MIMO applications.

## References

1. Rappaport TS et al (2013) Millimeter wave mobile communications for 5G cellular: it will work! IEEE Access 1:335–349. https://doi.org/10.1109/ACCESS.2013.2260813
2. Dybdal R (1983) Millimeter wave antenna technology. IEEE J Sel Areas Commun 1(4):633–644. https://doi.org/10.1109/JSAC.1983.1145968
3. Alhalabi RA, Rebeiz GM (2009) High-gain Yagi-Uda antennas for millimeter-wave switched-beam systems. IEEE Trans Antennas Propag 57(11):3672–3676. https://doi.org/10.1109/TAP.2009.2026666
4. Hussain MT, Sharawi MS, Podilchack S, Antar YMM (2016) Closely packed millimeter-wave MIMO antenna arrays with dielectric resonator elements. In: 2016 10th European conference on antennas and propagation (EuCAP), pp 1–4. https://doi.org/10.1109/EuCAP.2016.7481202
5. Ali MMM, Sebak AR (2016) Design of compact millimeter wave massive MIMO dual-band (28/38 GHz) antenna array for future 5G communication systems. In: 2016 17th international symposium on antenna technology and applied electromagnetics (ANTEM), pp 1–2. https://doi.org/10.1109/ANTEM.2016.7550213
6. Hagras A, Denidni TA, Nedil M, Coulibaly Y (2012) Low-mutual coupling antenna array for millimeter-wave MIMO applications. In: Proceedings of the 2012 IEEE international symposium on antennas and propagation, pp 1–2. https://doi.org/10.1109/APS.2012.6348094
7. Ahmad W, Ali A, Khan WT (2016) Small form factor PIFA antenna design at 28 GHz for 5G applications. In: 2016 IEEE international symposium on antennas and propagation (APSURSI), pp 1747–1748. https://doi.org/10.1109/APS.2016.7696580
8. Psychoudakis D, Wang Z, Aryanfar F (2013) Dipole array for mm-wave mobile applications. In: 2013 IEEE antennas and propagation society international symposium (APSURSI), pp 660–661. https://doi.org/10.1109/APS.2013.6710990
9. Dadgarpour A, Zarghooni B, Virdee BS, Denidni TA (2016) One- and two-dimensional beam-switching antenna for millimeter-wave MIMO applications. IEEE Trans Antennas Propag 64(2):564–573. https://doi.org/10.1109/TAP.2015.2508478
10. Dadgarpour A, Sorkherizi MS, Kishk AA, Denidni TA (2016) Single-element antenna loaded with artificial mu-near-zero structure for 60 GHz MIMO applications. IEEE Trans Antennas Propag 64(12):5012–5019. https://doi.org/10.1109/TAP.2016.2618485
11. Briqech Z, Sebak AR, Denidni TA (2016) Wide-scan MSC-AFTSA array-fed grooved spherical lens antenna for millimeter-wave MIMO applications. IEEE Trans Antennas Propag 64(7):2971–2980. https://doi.org/10.1109/TAP.2016.2565704
12. Ho CH et al (2013) Performance evaluation of a 60 GHz radio-over-fiber system employing MIMO and OFDM modulation. IEEE J Sel Areas Commun 31(12):780–787. https://doi.org/10.1109/JSAC.2013.SUP2.12130010

13. Zeng Y, Zhang R, Chen ZN (2014) Electromagnetic lens-focusing antenna enabled massive MIMO: performance improvement and cost reduction. IEEE J Sel Areas Commun 32(6):1194–1206. https://doi.org/10.1109/JSAC.2014.2328151
14. Zeng Y, Zhang R (2016) Millimeter wave MIMO with lens antenna array: a new path division multiplexing paradigm. IEEE Trans Commun 64(4):1557–1571. https://doi.org/10.1109/TCOMM.2016.2533490
15. Hong W, Baek K, Lee Y, Kim YG (2014) Design and analysis of a low-profile 28 GHz beam steering antenna solution for future 5G cellular applications. In: 2014 IEEE MTT-S international microwave symposium (IMS2014), pp 1–4. https://doi.org/10.1109/MWSYM.2014.6848377
16. Zhou H, Aryanfar F (2013) Millimeter-wave open ended SIW antenna with wide beam coverage. In: 2013 IEEE antennas and propagation society international symposium (APSURSI), pp 658–659. https://doi.org/10.1109/APS.2013.6710989
17. Dadgarpour A, Sorkherizi MS, Kishk AA (2016) Wideband low-loss magnetoelectric dipole antenna for 5G wireless network with gain enhancement using meta lens and gap waveguide technology feeding. IEEE Trans Antennas Propag 64(12):5094–5101. https://doi.org/10.1109/TAP.2016.2620522
18. Dadgarpour A, Zarghooni B, Virdee BS, Denidni TA (2015) Beam-deflection using gradient refractive-index media for 60-GHz end-fire antenna. IEEE Trans Antennas Propag 63(8):3768–3774. https://doi.org/10.1109/TAP.2015.2438396
19. Sharawi MS, Podilchak SK, Hussain MT, Antar YMM (2017) Dielectric resonator based MIMO antenna system enabling millimetre-wave mobile devices. Antennas Propag IET Microw 11(2):287–293. https://doi.org/10.1049/iet-map.2016.0457
20. Gupta S, Briqech Z, Sebak AR, Denidni TA (2017) Mutual-coupling reduction using metasurface corrugations for 28 GHz MIMO applications. IEEE Antennas Wirel Propag Lett 16:2763–2766. https://doi.org/10.1109/LAWP.2017.2745050
21. Al-Hasan MJ, Denidni TA, Sebak AR (2015) Millimeter-wave compact EBG structure for mutual coupling reduction applications. IEEE Trans Antennas Propag 63(2):823–828. https://doi.org/10.1109/TAP.2014.2381229
22. Bait-Suwailam MM, Boybay MS, Ramahi OM (2010) Electromagnetic coupling reduction in high-profile monopole antennas using single-negative magnetic metamaterials for MIMO applications. IEEE Trans Antennas Propag 58(9):2894–2902. https://doi.org/10.1109/TAP.2010.2052560
23. Sharawi MS, Numan AB, Aloi DN (2013) Isolation improvement in a dual-band dual-element MIMO antenna system using capacitively loaded loops. Prog Electromagn Res 134:247–266. https://doi.org/10.2528/PIER12090610
24. Szabo Z, Park G-H, Hedge R, Li E-P (2010) A unique extraction of metamaterial parameters based on Kramers-Kronig relationship. IEEE Trans Microw Theory Tech 58(10):2646–2653. https://doi.org/10.1109/TMTT.2010.2065310
25. Wani Z, Abegaonkar MP, Koul SK (2018) A 28-GHz antenna for 5G MIMO applications. Prog Electromagnet Res Lett 78:73–79. https://doi.org/10.2528/PIERL18070303
26. Wani Z, Vishwakarma DK (2016) An ultrawideband antenna for portable MIMO terminals. Microw Opt Technol Lett 58(1):51–57. https://doi.org/10.1002/mop.29498
27. Vaughan RG, Andersen JB (1987) Antenna diversity in mobile communications. IEEE Trans Veh Technol 36(4):149–172. https://doi.org/10.1109/T-VT.1987.24115
28. Sharawi MS (2017) Current misuses and future prospects for printed multiple-input, multiple-output antenna systems [wireless corner]. IEEE Antennas Propag Mag 59(2):162–170. https://doi.org/10.1109/MAP.2017.2658346
29. Tian R, Lau BK, Ying Z (2011) Multiplexing efficiency of MIMO antennas. IEEE Antennas Wirel Propag Lett 10:183–186. https://doi.org/10.1109/LAWP.2011.2125773
30. Hsu YW, Huang TC, Lin HS, Lin YC (2017) Dual-polarized quasi Yagi-Uda antennas with endfire radiation for millimeter-wave MIMO terminals. IEEE Trans Antennas Propag 65(12):6282–6289. https://doi.org/10.1109/TAP.2017.2734238

# Chapter 7
# Millimeter Wave Polarization/Pattern Diversity Antennas

## 7.1 Introduction

The antenna diversity is very effective for improving the mobile to base station communication link quality. The idea behind the antenna diversity is to have multiple antenna outputs that are not correlated; thus, it becomes highly unlikely that all the output signals are in deep fade. Multiple antenna diversity dimensions that are available are spatial diversity, polarization diversity and pattern diversity. In spatial diversity, antennas are separated by physical distance to have un-correlated channel matrix. For polarization and pattern diversity, multiple antennas have different polarizations and directional radiation beams along different angular space respectively. The polarization diversity has been found as effective as spatial diversity in which antennas are separated by about half wavelength apart, therefore implementing polarization diversity with pattern diversity can save good amount of antenna space [1]. Compared to sub-6 GHz wireless communications, the mmWave 5G communication would use highly directional antennas and maintaining the link with randomly oriented mobile devices would require mmWave MIMO antennas with pattern and polarization diversity.

This chapter focusses on the design of millimetre wave antennas exploiting pattern and polarization diversity techniques for better coverage and reception of the mmWave wireless signals.

## 7.2 Antenna Polarization and Pattern Diversity

The mmWave antennas with diversity provide major benefits in terms of reliability and reduction in outage probability. For example, a polarization mismatch between transmitter and receiver can reduce the signal power by 10–20 dB. Antennas with the polarization and pattern diversity can automatically adjust the antenna radiation



S. K. Koul and Z. Wani, *Novel Millimetre Wave Antennas for MIMO and 5G Applications*, Lecture Notes in Electrical Engineering 819,
https://doi.org/10.1007/978-981-16-7278-1_7

characteristics to receive maximum signal power, thereby increasing the link SNR. The diversity antennas have the flexibility in terms of choosing or mixing the output signals for better SNR. The output signal can be selected based on higher SNR or can be mixed after co-phased weighting [2].

### 7.2.1 Dual-Polarized mmWave Antenna

A fully integrated dual-polarized antenna with Yagi-Uda antenna elements is shown in Fig. 7.1. The antenna consists of horizontal and vertically polarized antennas implemented on a three-layer PCB substrate. The vertical polarization is achieved using three vias in substrate acting as radiating element and two directors each measuring quarter wavelength in height as shown in the cross-sectional view in Fig. 7.1b. To push the radiation towards the end-fire direction and increasing front-to-back ratio, a corner shaped reflector made using vias behind the radiator is used. The horizontal polarization is achieved for port 2, which consists of classic quasi-Yagi-Uda antenna with two directors as shown in Fig. 7.1c. Both the ports (1 and 2) have orthogonal polarization diversity and each port radiates in end-fire radiation. Both these ports cover 33–40 GHz band with an isolation greater than 20 dB resulting due to the orthogonal polarization of the two antennas as shown in Fig. 7.1d. The design is then extended further to generate more directed radiation beams using $1 \times 4$antenna array to achieve a peak gain of 11.8 dBi.

The MIMO performance of the two-port dual-polarized can be evaluated using the performance matrix known as cross polarization discrimination (XPD). The XPD represents the polarization purity of the vertically and horizontal radiation beams. For a $2 \times 2$ MIMO channel with horizontal and vertical polarized antennas, the channel matrix can be written as [4]:

$$H = \begin{bmatrix} H_{11} & H_{12} \\ H_{21} & H_{22} \end{bmatrix} \tag{7.1}$$

where,

$H_{11} = \begin{bmatrix} h_{1V,1V} & h_{1V,1H} \\ h_{1H,1V} & h_{1H,1H} \end{bmatrix}$, $H_{12} = \begin{bmatrix} h_{1V,2V} & h_{1V,2H} \\ h_{1H,2V} & h_{1H,2H} \end{bmatrix}$, $H_{21} = \begin{bmatrix} h_{2V,1V} & h_{2V,1H} \\ h_{2H,1V} & h_{2H,1H} \end{bmatrix}$ and $H_{22} = \begin{bmatrix} h_{2V,2V} & h_{2V,2H} \\ h_{2H,2V} & h_{2H,2H} \end{bmatrix}$.

The $h_{iV,jV}$ and $h_{iH,jH}$ are the scaler channel parameters between $i$th antenna and $j$th antenna for vertical and horizontal polarizations, respectively. The cross-polarization components are denoted by $h_{iV,jH}$ and $h_{iH,jV}$. The XPD for the $2 \times 2$ MIMO with H-pol and V-pol components is given by:

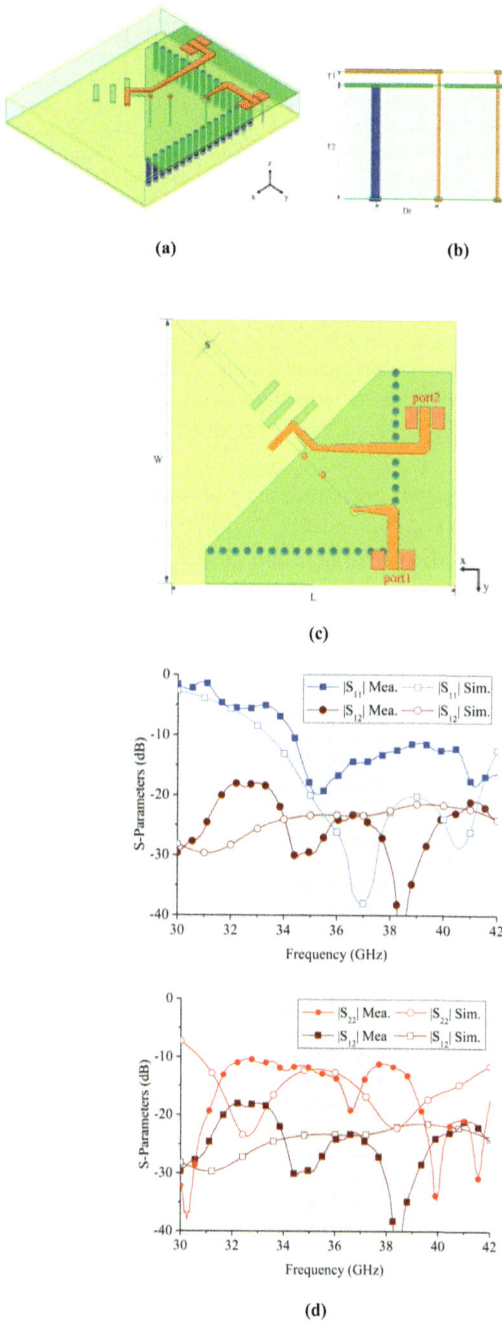


**Fig. 7.1** Two-port dual-polarized antenna **a** perspective view, **b** cross-sectional view, **c** top view and **d** measured and simulated s-parameters of the antenna ([3], Reprinted with permission from IEEE)

$$XPD = \frac{E\{|h_{H,H}|^2\}}{E\{|h_{V,H}|^2\}} = \frac{E\{|h_{V,V}|^2\}}{E\{|h_{H,V}|^2\}} \tag{7.2}$$

### 7.2.2 mmWave Antenna with Pattern Diversity

The high spectral efficiency achieved using spatial multiplexing technique takes a hit on signal reliability in case of mmWave MIMO communications as the communication links are power limited. Therefore, antennas with higher beamforming gains and reconfigurable radiation beams to suit the conditions of the wireless channel are essential. The pattern diversity of such antennas results in improved system performance.

A mmWave antenna with two ports radiating orthogonally is shown below in Fig. 7.2. The antennas consist of two end-fire Yagi-Uda antennas with five directors each on thin substrate with dielectric constant of 2.2, operating at 28 GHz. The schematic of the antenna consists of two Yagi-Uda antennas for landscape and portrait modes of use case shown in Fig. 7.2. Both these antennas radiate in two orthogonal end-fire directions for maintaining the signal reception, if one of the beams is blocked by the human hand in one of the modes of operation. Each antenna has a forward gain of 9 dBi essential for compensating link loss and beamforming to orthogonal directions can be done using a SPDT switch. Since both these antennas are bend at 90°, we expect there is hardly any mutual coupling between the two ports which is quite essential for an efficient MIMO system. The mutual coupling between the two antennas is presented in Fig. 7.3a. The two radiation patterns $\phi = 180°$ and 270° are shown in Fig. 7.3b.

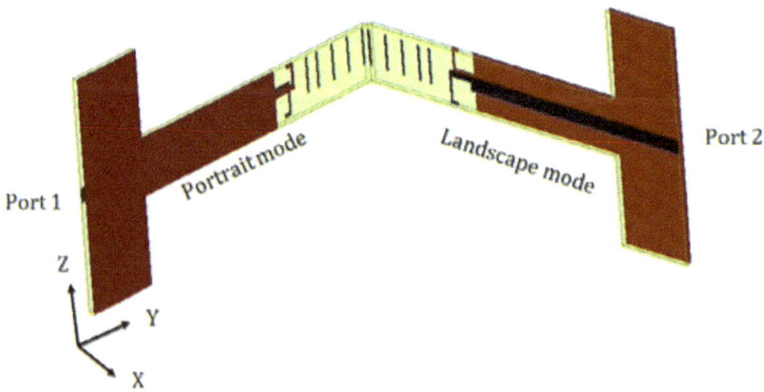


**Fig. 7.2** Schematic of the two port antennas with orthogonal pattern diversity ([5], Reprinted with permission from Wiley)

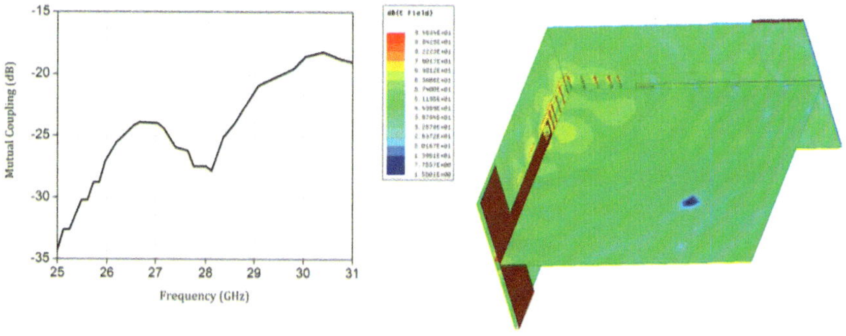


(a)

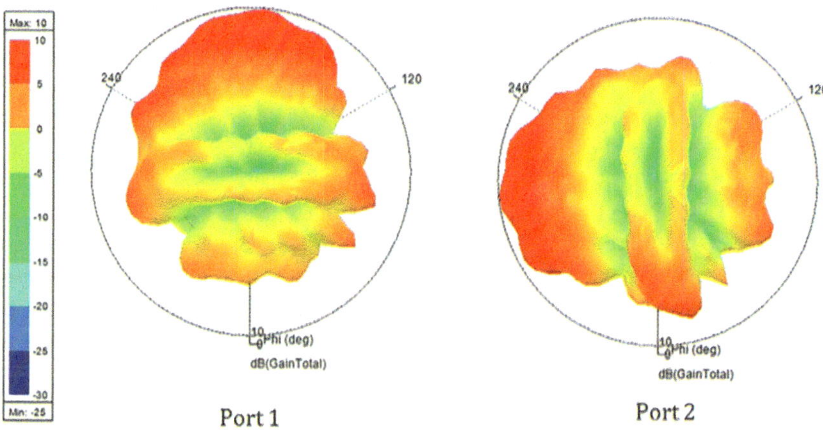


(b)

**Fig. 7.3** **a** Mutual coupling and E-filed distribution depicting coupling, **b** 3D radiation pattern for two ports ([5], Reprinted with permission from Wiley)

## 7.3 Compact mmWave Antenna with Pattern and Polarization Diversity

The mutual coupling between the antennas in MIMO configuration plays an important role in deciding the diversity/multiplexing performance of the antenna system. There are various techniques to improve the isolation between antennas e.g., using artificial media, parasitic elements, and band gap structures [6]. A two-port shared aperture antenna with dual-polarized zero index metamaterials in [7] covers 27–30 GHz band with an isolation of 22 dB between two ports and has an average gain of 9 dBi for each port packed in an overall size of 31 mm × 31 mm. To achieve similar performance in terms of gain, isolation, and pattern/polarization diversity at much smaller physical footprint, two dipole antennas placed very close to each other with a dual-polarized high index surface is designed as follows.

### 7.3.1 Shared-Aperture Antenna

The design of shared aperture MIMO antenna consists of two end-fire antennas radiating through a single high index surface for gain improvement. The design of two-port antenna and dual-polarized HIS is presented. Microstrip fed dipole antennas are designed on a Rogers substrate ($h = 0.254$ mm, $\epsilon_r = 2.2$) and are placed orthogonal to each other as shown in Fig. 7.4a. The two dipole radiators are packed in compact physical footprint and are separated by just 1.4 mm, however this results in degraded isolation between the two ports. The simulated S-parameters of the antenna-1 (same for antenna-2) are shown in Fig. 7.4b. The isolation between the antennas ranges from 13 to 16 dB in the 25–30 GHz band and both these antennas radiate and have a 6 dBi peak directivity over the band. The large mutual coupling between the antennas results due to the coupling of the electric field components in the two radiators placed very close to each other. Since the antenna has a very thin substrate with lower permittivity values, most of the fields are in air than in the substrate. The radiated electric field from antenna-1 excites the radiator-2, which results in coupling of the fields to other port. The snapshot of the electric fields for ports 1 and 2 are shown in Fig. 7.5. We can observe that the isolation is degraded due to the coupling of near-field components of the electric field.

As can be seen, the gain of the two antennas is not more than 6 dB, the gain can be enhanced using the integration of the metamaterial unit cell array on the shared substrate space between the two radiators. Typically, the gain enhancement in case

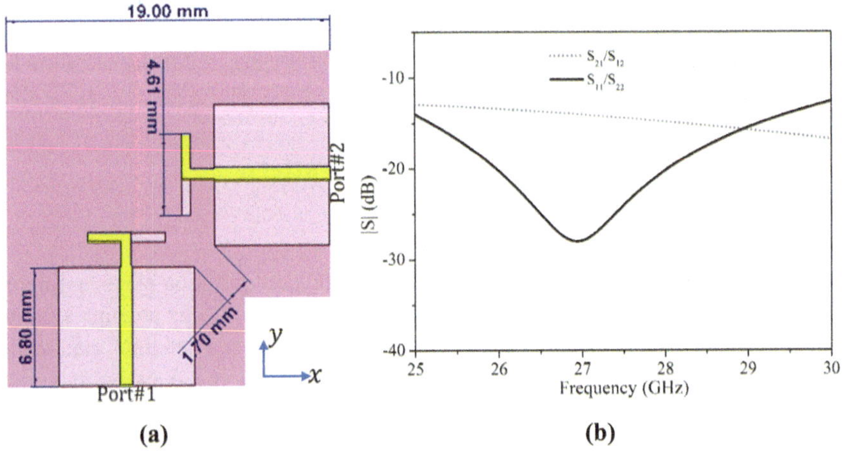


**Fig. 7.4** **a** Configuration of the shared aperture antenna and **b** two port S-parameters

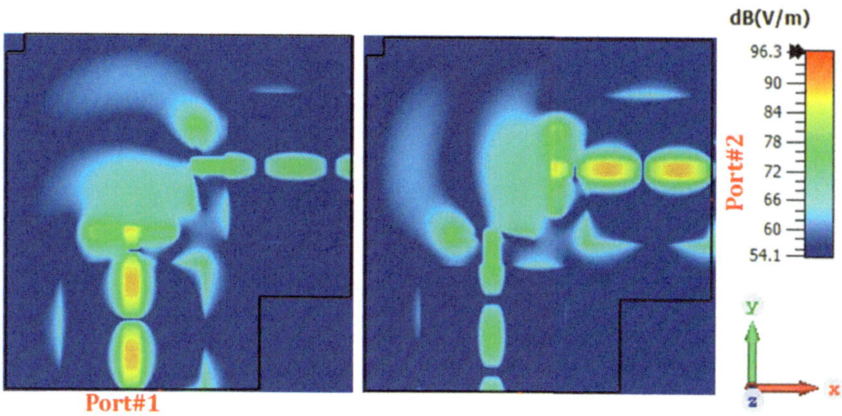


**Fig. 7.5** Snapshot of the near-field electric field depicting the mutual coupling when port 1 and port 2 are active

of end-fire antennas is achieved using directors in front of the radiator, however such technique is polarization sensitive. Here the gain improvement for both the ports is done using dual-polarized metamaterial unit cell array. The gain of the two dipole antennas is enhanced by guiding the TE surface waves without phase accumulation along the two orthogonal directions of propagation. Since the two dipoles are printed on very thin substrate having a refractive index of 1.4, the power in TE surface waves in not much. The power in TE surface waves can be increased using the high refractive index surface (HIS) generated using an array of non-resonant metamaterial unit cells. To generate a high index surface, a dual-polarized metamaterial unit cell is designed as shown in Fig. 7.6 having same properties for the two orthogonal orientations. The refractive index of the substrate can be increased to 2.1 by printing array of such unit cells. A high refractive index can be achieved using a non-resonant metamaterial unit cells [8]. As shown in Fig. 7.6b, the unit cell is operating in a non-resonant operation for the 25–30 GHz band, thereby high permittivity is obtained resulting in a high refractive index for the band. As seen from the figure, the refractive index of the unit cell on the same substrate is higher than the substrate only case. These parameters are extracted using the PEC boundary along x/y-direction and PMC boundary along z-direction with excitation along y/x-direction depicting the actual field configuration for the two dipole antennas [9]. Since the unit cell is symmetrical for the two orientations and thus have similar properties for $k_x$ and $k_y$ vectors.

### 7.3.2 *Shared Aperture Antenna with Dual-Polarized HIS*

The configuration of the two-port shared aperture antenna with high index surface is shown in Fig. 7.7. By printing the dual-polarized unit cells on the shared aperture

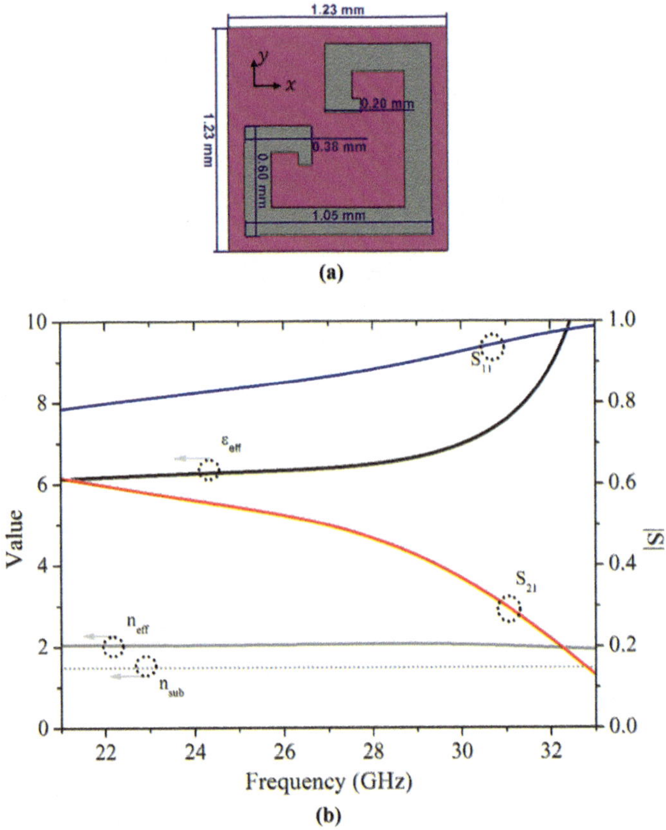


**Fig. 7.6** HIS unit cell **a** geometrical configuration and **b** extracted parameters of the unit cell from the S-parameters

substrate, the TE surface wave can be contained and focused along the two orthogonal directions resulting in gain improvement for the two antennas firing the same shared aperture HIS [10].

The surface waves radiated from the antennas along the two orthogonal directions are guided without any phase accumulation and which eventually enhances the directivity of the antennas. To get an insight into the directivity enhancement by printing the HIS over the shared aperture substrate, the tangential component of the radiated electric field is plotted along x- and y-directions in Fig. 7.8. The magnitude of the radiated electric field with HIS is compared with the electric field without HIS. It is observed that the electric field is more focused along the direction of propagation by HIS resulting in higher directivity.

The 3D radiation pattern of the aperture coupled antenna with the HIS is compared with antenna without HIS in Fig. 7.9. As can be seen that when HIS is not present, some power is radiated along the z-direction which eventually results in lower gain.

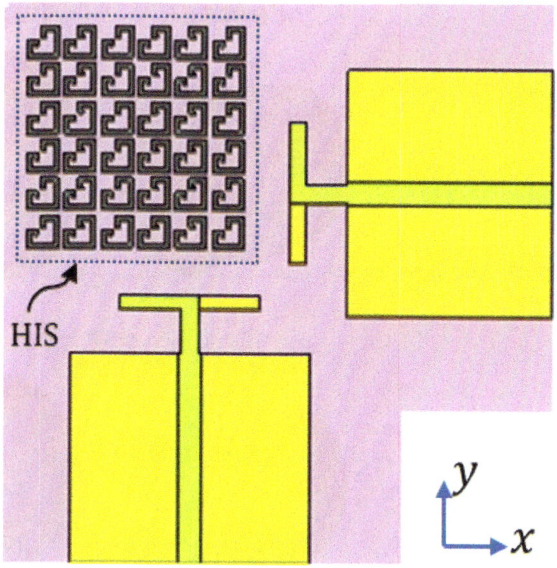


**Fig. 7.7** Configuration of the two-port shared aperture antenna with HIS

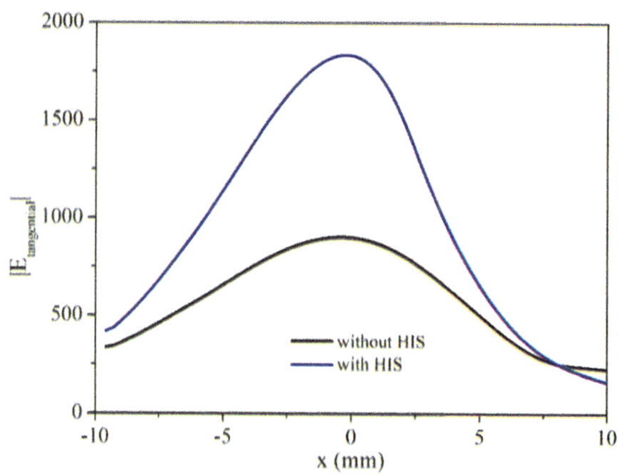


**Fig. 7.8** Magnitude of radiated electric field along x-direction for port 1 (same for port 2)

Also, we can observe that radiation for both the ports is focused along end-fire directions when HIS is present. This is due to the coupling of radiated fields with the HIS and guiding along the end-fire direction.

The two-port shared aperture antenna with HIS covers the 25–32 GHz band, but still suffers with a lower port isolation of 14 dB at the 28 GHz band. There is no change in the mutual coupling between the two when HIS is printed. Now the tangential components of the electric fields from antenna 1 couples back to antenna 2 through the unit cell shown as inset in Fig. 7.10. To stop the fields coupling back to the other ports through HIS, a totally reflecting metamaterial unit cells are printed

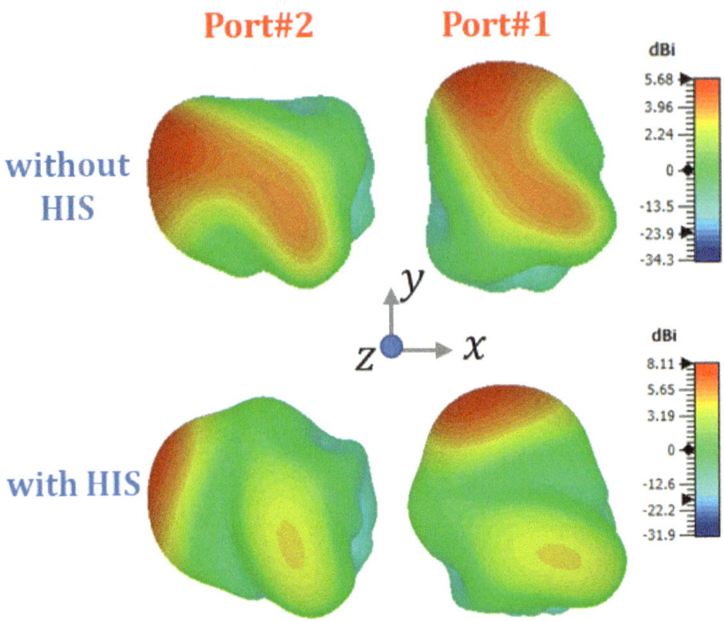


**Fig. 7.9.** 3D radiation pattern of the two-port shared aperture antenna with and without HIS at 28 GHz

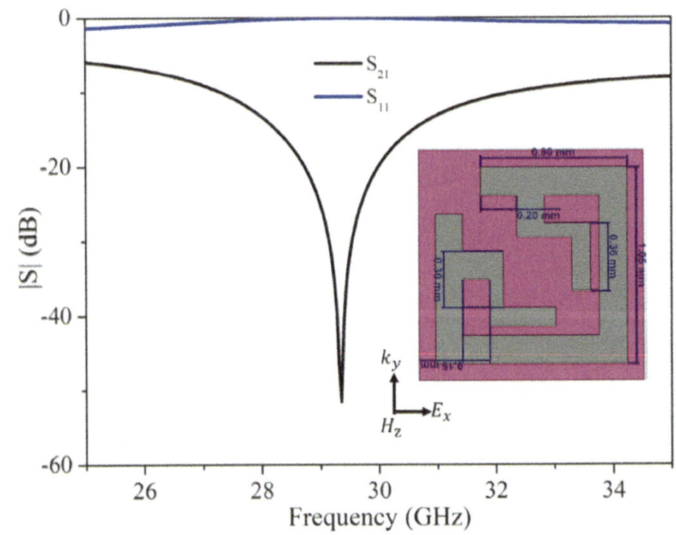


**Fig. 7.10** S-parameters of the totally reflect unit cell along with the geometry of the unit cell

on back between the radiators and HIS. The totally reflect unit cell is modified dual-polarized HIS unit cell. The resonant frequency is now tuned to the band of interest to have no transmission for the 25–30 GHz band. The tuning is done by increasing length of the metallic loops. The S-parameters of the modified unit cell for the $E_x$ and $H_z$ field configuration is shown in Fig. 7.10. The unit cell has a $|S_{21}|$ dip of -55 dB at around 29 GHz and acts as a total reflector for the x- and y-polarized waves required to stop the fields from antenna-1 to antenna-2 and vice-versa. Array of $5 \times 1$ such unit cells are printed on bottom of the substrate between the two radiators and the HIS. These unit cells stop the coupling of the fields form one port to other port through HIS and results in an isolation of 20 dB at 28 GHz as against of 14 dB without these metamaterial unit cells.

## 7.4 Prototype and Measurement Results

The two-port shared aperture antenna with HIS and improved isolation using total reflect unit cells printed on back side is shown in Fig. 7.11a. The antenna is fabricated using the photolithography process on a Rogers substrate. Two 2.92 mm end-launch connects are used for measuring the port performance of the antenna using a Keysight PNA.

The measured and simulated S-parameters of the shared aperture prototype antenna are plotted in Fig. 7.11b. It is observed that shared aperture antenna has a reflection coefficient $|S_{11}| < -10$ dB over 25–29 GHz frequency band. Also, the impedance bandwidth is not disturbed by the printed HIS and totally reflect unit cells. The mutual coupling between the antennas has been reduced using the modified metamaterial unit cells, measured isolation greater than 18 dB is achieved across the band.

The radiation characteristics of the two-port shared aperture antenna is measured in a far-field type anechoic chamber using standard Ka-band horn antenna. The gain measurements performed using a pair of Ka-band horn antennas and protype antenna are plotted and compared with the antenna gain without HIS loading in Fig. 7.12. The results show a peak gain of 8.68 dB at around 28.5 GHz with an overall 4 dB of enhancement with respect to no-HIS loading. The simulated and measured normalized radiation patterns are plotted for ports 1 and 2 in Fig. 7.13 at 28 and 29 GHz. The radiation patterns are titled towards the HIS due to offset position of the HIS with respect to radiators which results in radiation beams pointing along $\phi = 110°$ and $160°$. As a result, pattern diversity is achieved with orthogonal polarizations for ports 1 and 2.

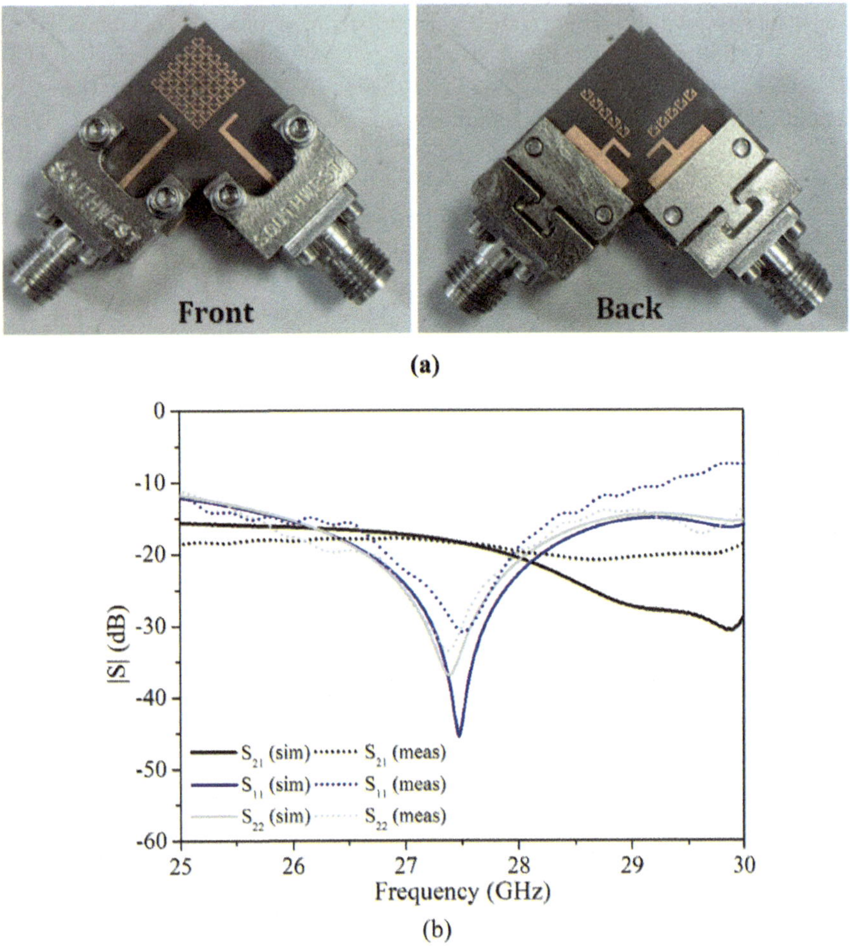


**Fig. 7.11** **a** Photograph of the prototype shared aperture antenna and **b** measured and simulated scattering parameters of the antenna

## 7.5 Conclusion

This chapter presents mmWave antennas with pattern and polarization diversity for wireless communications. A dual-polarized mmWave antenna integrated in a three-layer substrate is presented. The fully integrated dual-polarized antenna can be a potential candidate for handheld mobile terminals with its corner antenna design approach. Later shared aperture antenna with pattern diversity is presented for mmWave communication. The pattern diversity is very crucial for mmWave communication for link reliability. The shared aperture antenna covers 28 GHz band with two orthogonal radiation beams having 9 dBi peak gains.

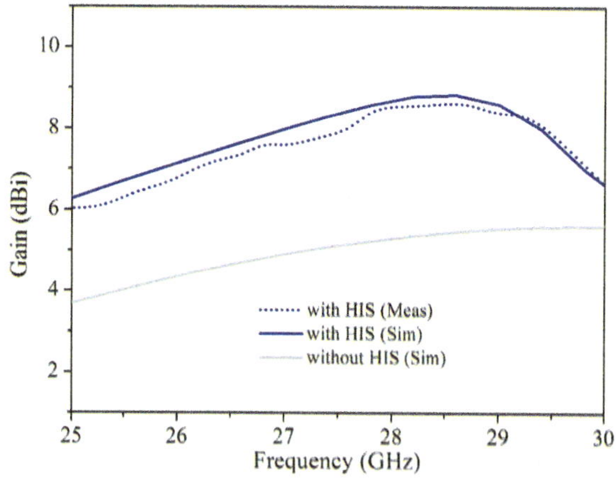


**Fig. 7.12** Simulated and measured gain of the prototype antenna at port-2 compared with the gain of the antenna without HIS loading

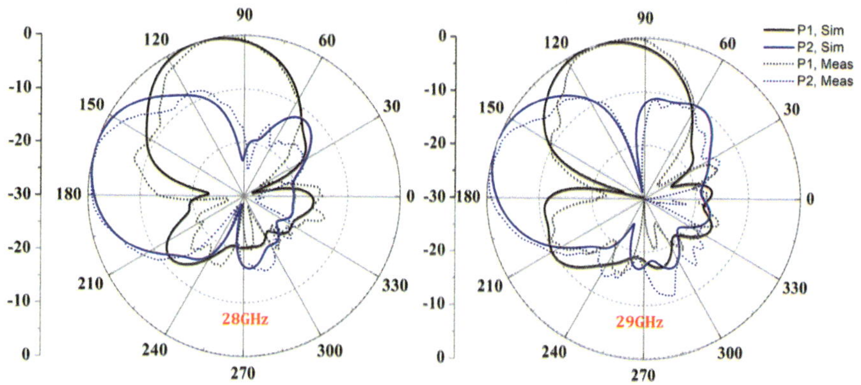


**Fig. 7.13** Measured and simulated normalized E-plane radiation patterns for ports 1 and 2 at 28 and 29 GHz

## References


1. Lee W, Yeh Y (1972) Polarization diversity system for mobile radio. IEEE Trans Commun 20(5):912–923. https://doi.org/10.1109/TCOM.1972.1091263
2. Jakes WC (1974) Diversity techniques. In: Microwave mobile communications. IEEE, pp 389–544. https://doi.org/10.1109/9780470545287.ch6
3. Hsu YW, Huang TC, Lin HS, Lin YC (2017) Dual-polarized quasi Yagi-Uda antennas with endfire radiation for millimeter-wave MIMO terminals. IEEE Trans Antennas Propag 65(12):6282–6289. https://doi.org/10.1109/TAP.2017.2734238
4. Coldrey M (2008) Modeling and capacity of polarized MIMO channels. In: VTC spring 2008—IEEE vehicular technology conference, pp 440–444. https://doi.org/10.1109/VETECS.2008.103

5. Karthikeya GS, Koul SK, Poddar AK, Rohde U (2021) Ultra-compact orthogonal pattern diversity antenna module for 5G smartphones. Microw Opt Technol Lett 63(8):2003–2012. https://doi.org/10.1002/mop.32378
6. Khajeh-Khalili F, Honarvar MA, Naser-Moghadasi M, Dolatshahi M (2020) Gain enhancement and mutual coupling reduction of multiple-input multiple-output antenna for millimeter-wave applications using two types of novel metamaterial structures. Int J RF Microw Comput Aided Eng 30(1):e22006. https://doi.org/10.1002/mmce.22006
7. Gulur Sadananda K, Abegaonkar MP, Koul SK (2019) Gain equalized shared-aperture antenna using dual-polarized ZIM for mmWave 5G base stations. IEEE Antennas Wirel Propag Lett 18(6):1100–1104. https://doi.org/10.1109/LAWP.2019.2910183
8. Wani Z, Abegaonkar MP, Koul SK (2020) High-low-epsilon biaxial anisotropic lens for enhanced gain and aperture efficiency of a linearly polarized antenna. IEEE Trans Antennas Propag 68(12):8133–8138. https://doi.org/10.1109/TAP.2020.2983787
9. Chen X, Grzegorczyk TM, Wu B-I, Pacheco J, Kong JA (2004) Robust method to retrieve the constitutive effective parameters of metamaterials. Phys Rev E 70(1):016608. https://doi.org/10.1103/PhysRevE.70.016608
10. Kaneda N, Deal WR, Qian Y, Waterhouse R, Itoh T (2002) A broadband planar quasi-Yagi antenna. IEEE Trans Antennas Propag 50(8):1158–1160. https://doi.org/10.1109/TAP.2002.801299

# Chapter 8
# Dual Beam/Multibeam Antennas

## 8.1 Introduction

The use of MIMO antennas at mmWave frequencies help in achieving higher communication throughput and helps in maintaining the wireless link. As discussed in Chap. 6, the MIMO with beamforming can have much better performance than traditional MIMO techniques at mmWaves. Multibeam antennas generate multiple radiation beams to cover predefined angular space and provide added advantage as compared to single beam antennas [1]. Furthermore, the dual/multi beam antennas would cover a large area with a reduced number of antennas. Also, the 28 GHz communication is severely affected by the interference and fading. Dual/multi beam antennas along with the MIMO would overcome these issues with improved communication link quality [2, 3]. At millimetre wave frequencies, several antennas with MIMO capability have been proposed but, each antenna radiates a single beam [4, 5]. A study presented in [6] showed effect of combining bi-beam, tri-beam and quad-beam on path loss and path shadowing. Authors showed that beam combining can improve signal quality, path loss and shadow fading, which results in better signal coverage.

The antennas presented so far had single radiation beam only, however studies at mmWaves have showed the performance of dual beam MIMO and the classical MIMO. It has been shown that the dual beam MIMO has a power gain of 1.6 dB with increased spatial diversity as compared to the classical MIMO. This chapter presents dual beam/multibeam antennas for better reception of the signals at mmWave frequencies where the signal power is limited due to various factors.

 
S. K. Koul and Z. Wani, *Novel Millimetre Wave Antennas for MIMO and 5G Applications*, Lecture Notes in Electrical Engineering 819,
https://doi.org/10.1007/978-981-16-7278-1_8

## 8.2 Dual-Beam Antenna Array

The design of dual-beam antennas at mmWave frequencies is very crucial for achieving higher communication throughput and link reliability. One of the simplest techniques to generate dual-beam radiation is using two antennas with equal magnitude but opposite phase currents. Based on this a dual-beam antenna at 27–33 GHz is designed using array of dipole antennas as follows.

The millimetre wave dual-beam antenna configuration is shown in Fig. 8.1. A 10 mil Neltec substrate having relative permittivity of 2.2 is used. The dipole antenna is optimized in CST Microwave Studio to cover 28 GHz frequency band and is fed using a 50 Ω transmission line. The dipole antenna radiates in end-fire direction with E-field in xy-plane. The feed lines are extended to connect the end-launch connectors for measurement purpose. Each port consists of two dipole antennas and their directors fed via a power divider. The dipole antennas are separated with a distance, $S_d$ = 6.6 mm, and the arrangement of dipole antennas is such that the fields radiated would cancel out along the bisector to the dipole array axis due to the equal path lengths from the two dipole radiators with opposite currents which results in dual-beam radiation for better diversity performance. After that the antennas are arranged orthogonally in MIMO configuration to achieve pattern diversity in the E-plane. The photograph of the fabricated antenna is shown in the Fig. 8.2a. For port performance measurement, high frequency connectors up to 40 GHz are used, the simulated and measured reflection coefficient along with mutual coupling between the two ports is shown in the Fig. 8.2b. The antenna covers 27–33 GHz frequency band with $|S_{11}| < -10$ dB and the coupling $|S_{21}| < -28$ dB, which is essential for

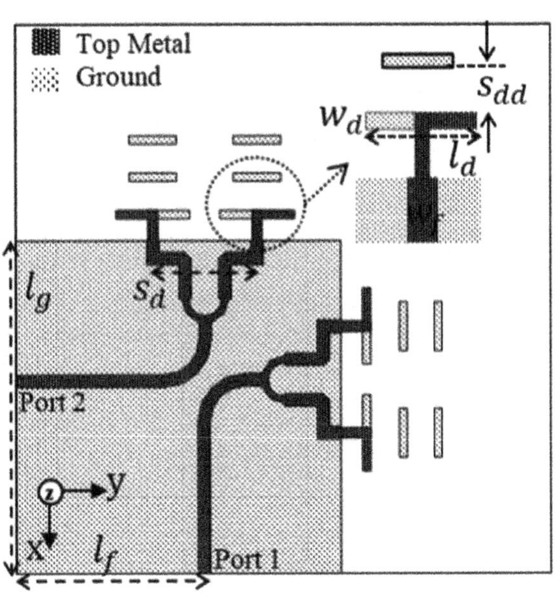


**Fig. 8.1** Dual-beam MIMO antenna geometry. ($S_{dd} = 1.8$, $w_f = 0.77$, $l_d = 4.7$, $w_d = 0.5$, $l_g = 20.6$, $l_f = 11.9$, $S_d = 6.6$, all dimensions are in millimetres). ([7], Reprinted with permission from IEEE)

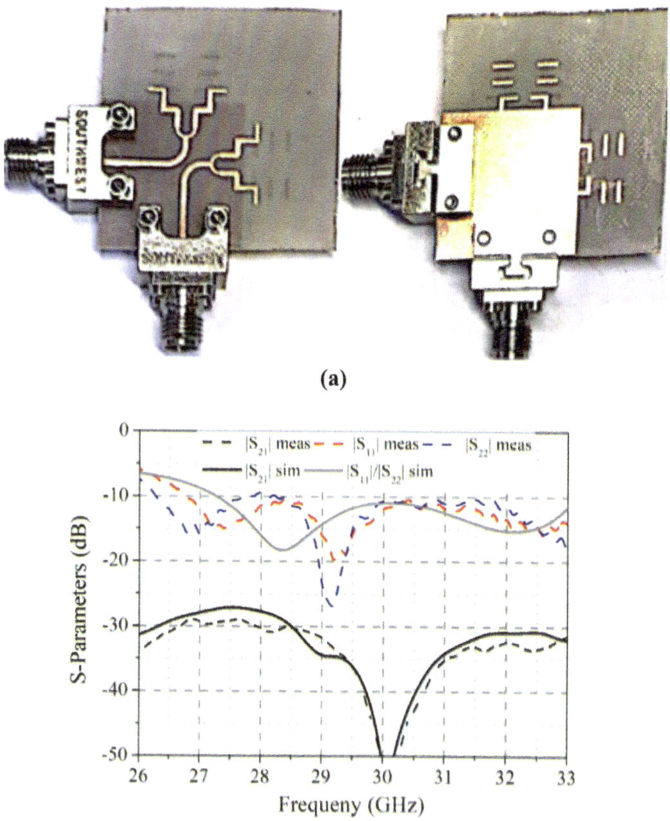


(b)

**Fig. 8.2** **a** Photograph of the fabricated prototype antenna, **b** measured and simulated S-parameters of the dual-beam MIMO antenna. ([7], Reprinted with permission from IEEE)

diversity performance. This high isolation is due to the orthogonal placement of the antennas with the edges of the connected ground plane between ports 1 and 2 which acts as resonator at 30 GHz, thus isolation more than 40 dB is observed from the figure.

In order to study the radiated fields from the two dipole antennas arrayed together, the antenna surface currents are shown in Fig. 8.3a. As can be seen from this Figure, opposite currents flow in the radiators. The fields from the two antennas would cancel along the central line to the array due to the equal path length from the two antennas, resulting in dual-beam radiation. The dual-beam radiation pattern of the antenna can be explained from the radiated E-field shown in the Fig. 8.3b. It is observed that the electric field emitted from the dipole antennas in array cancel out along the central line resulting in two beams for one port.

The 3D radiation pattern of the antenna for ports P1 and P2 is shown in Fig. 8.4 at

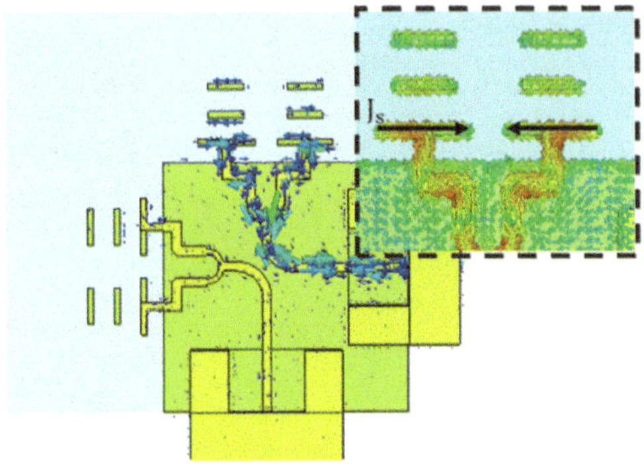


(a)

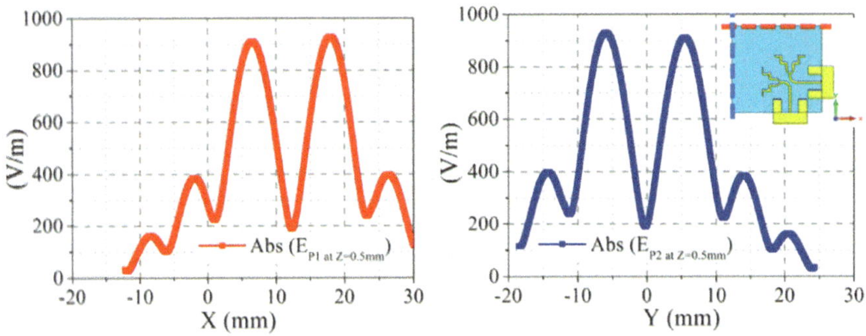


(b)

**Fig. 8.3** **a** The surface current distribution for port 1 and **b** E-field plot for ports-1 and 2 along X-axis and Y-axis for cut plane (Z = 0.5 mm, $\omega t = 0$)

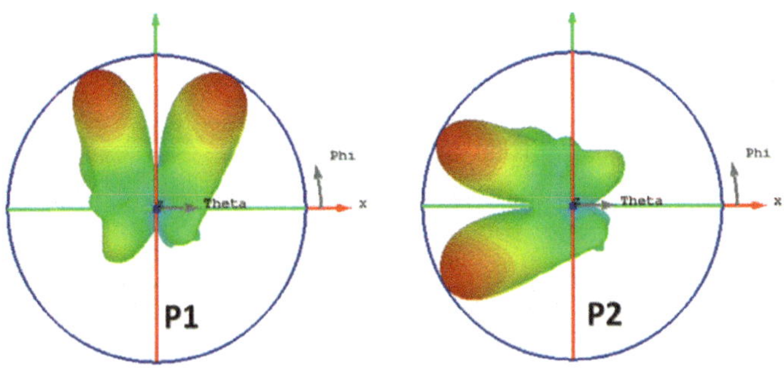


**Fig. 8.4.** 3D radiation pattern of the dual-beam antenna for ports P1 and P2 at 28 GHz

28 GHz. The radiation characteristics of the prototype dual-beam antenna is tested in a far-field type anechoic chamber. Figure 8.5 shows the measured and simulated E-plane pattern of the antenna at 27, 28.5 and 30 GHz. The radiation patterns are measured, and the results verified the predicted patterns of the dual-beam antenna. It is evident from Fig. 8.5, port 1 is having dual-beam radiation pattern with beams directed at 60° and 120° and for port 2, beams are directed at 150° and 210°, with side lobe levels less than − 11 dB. Figure 8.6 shows, each beam has a realized gain of more than 8 dBi and the gain is stable over the frequency band with peak gain of 9 dBi. The proposed antenna covers radiation angles from 45° to 225° with radiation nulls at 90° and 180°, with each radiation beam having 30° of beam-width in azimuthal plan. Normally the radiation nulls would result in transmission loss in LOS direction if any misalignment happens, however, the dual-beam MIMO doubles the probability of signal reception in a rich scattering environment by sending more radiation beams into the area.

## 8.3 Dual-Beam Radiation from a Single Antenna

Recently many researchers have proposed multi/dual beam antennas for millimeter wave applications [8–10]. Authors in [8], employed modified H-shaped resonators to enhance the effective refractive index on both sides of the Bowtie antenna to manipulate the TE surface waves to generate dual-beam radiation as shown in Fig. 8.7. However, there is about − 9 dB normalized radiation along the end-fire direction at lower frequency band which is not suitable for practical applications. Authors have used isotropic medium with high epsilon to generate dual-beam at 60 GHz when this medium loads the bow-tie antenna. The mechanism of dual-beam generation is established by exploiting the dielectric slab mode TE in the printed bow-tie dipole antenna. The authors have employed TE surface waves in the dielectric of the antenna to interact with high-permittivity medium employed using printed unit cells on the substrate. However, the power in surface wave modes in a substrate is very low when $h \ll \lambda_0/10$, where $h$ is the substrate height and $\lambda_0$ is free space wavelength [11]. Authors have used $h = 0.254$ mm (clearly, $h \ll 0.5$ at 60 GHz). It is clearly mentioned about the amount of power radiated in air and in TE surface mode for different dielectric constants from 1 to 16 in [11]. It is observed that for lower dielectric constant values (around 2) the radiated power is more than the TE surface waves power even when the dielectric height is large.

Therefore, with a 10 mil substrate having ($h = 0.254$ mm, and $\epsilon_r = 2.2$), the height $h$ of the substrate is very small as compared to ($\frac{\lambda_0}{10} = 1.071$) at 28 GHz, thus generating dual-beam based on TE surface mode in dielectric is not possible as the fields in free-space are prominent. Also, we have performed simulations based on the method described in [8] using a high permittivity substrate on both sides to interact only with the TE surface waves so as to generate dual-beam at 28 GHz. However, this is not possible as shown in the 3D pattern of the antenna in Fig. 8.8.

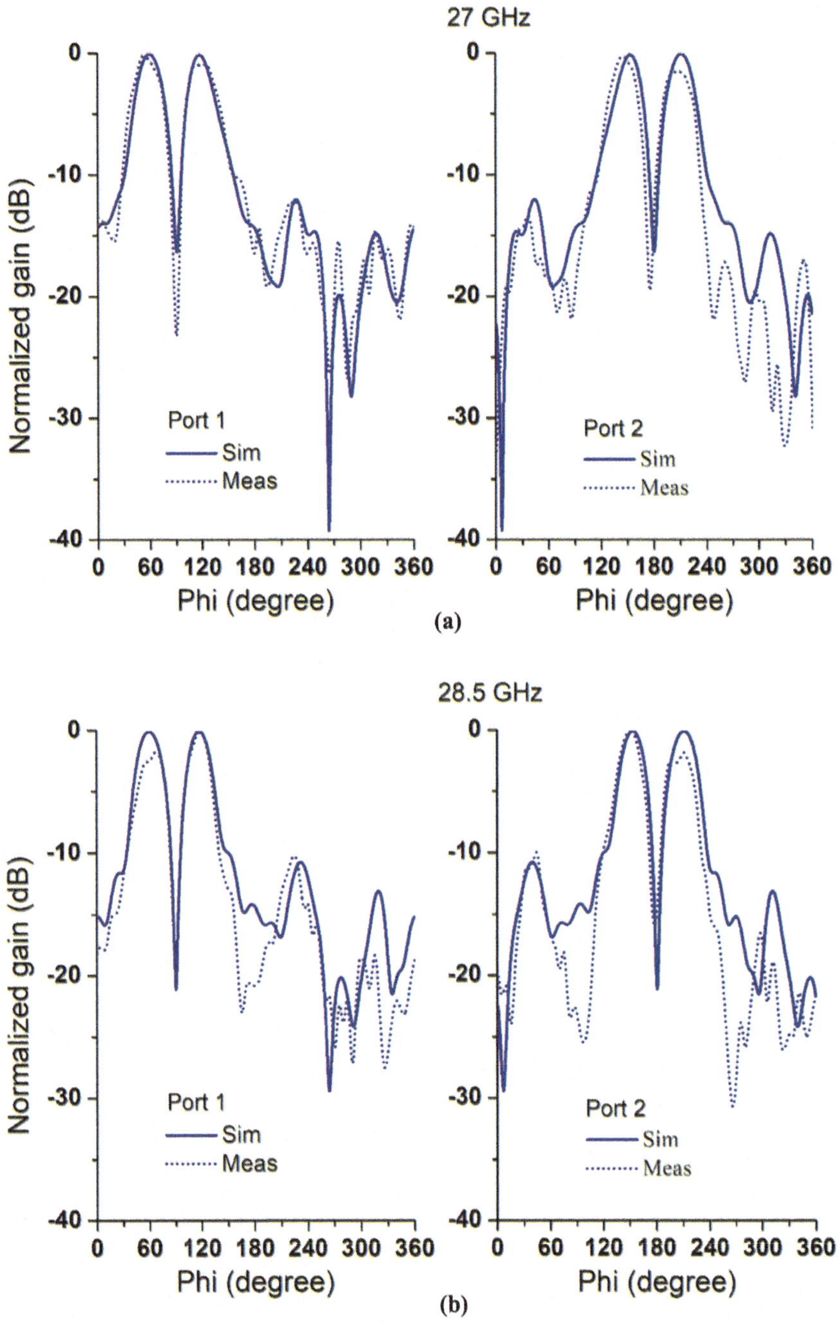


**Fig. 8.5** Measured and simulated E-plane radiation pattern of the proposed antenna when excited at two ports for **a** 27 GHz, **b** 28.5 GHz and **c** 30 GHz. ([7], Reprinted with permission from IEEE)

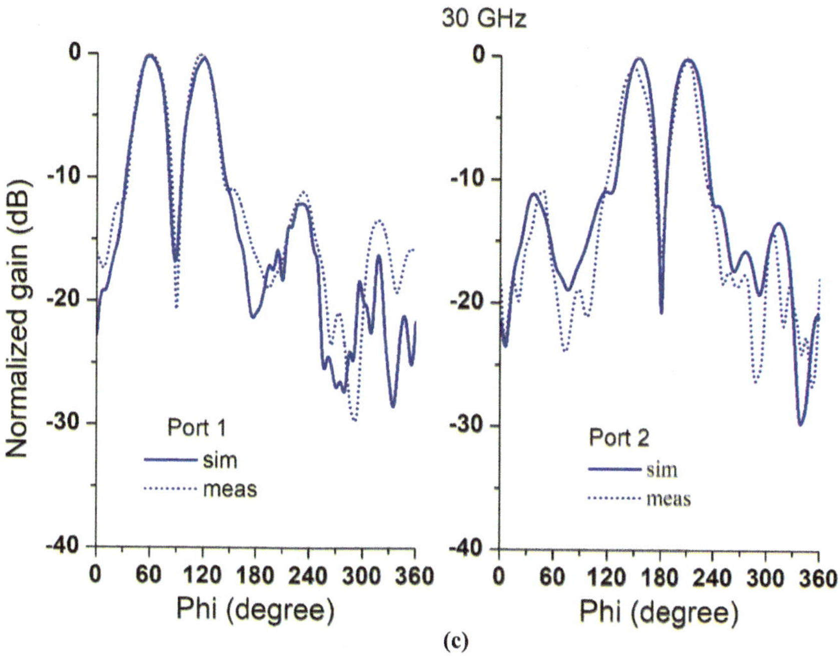


**Fig. 8.5** (continued)

**Fig. 8.6** Gain of the dual-beam MIMO antenna over the frequency band 26–33 GHz. ([7], Reprinted with permission from IEEE)

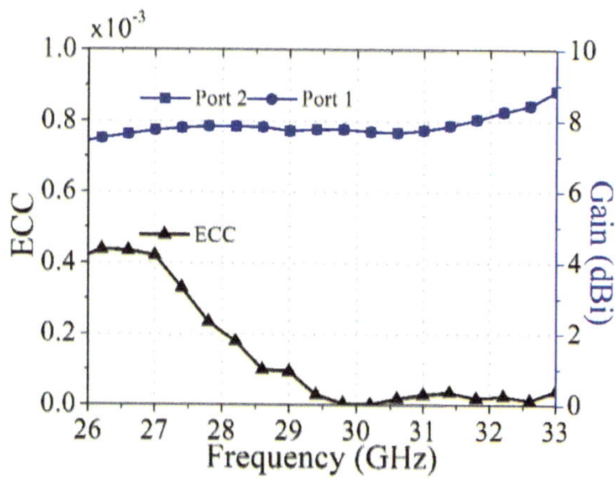


As this method considers TE surface waves in the substrate, it is required that the fields be in the substrate only neglecting the fields in free space. Thus, generating dual beam at 28 GHz with this thin substrate is not possible only with the isotropic high permittivity dielectric.

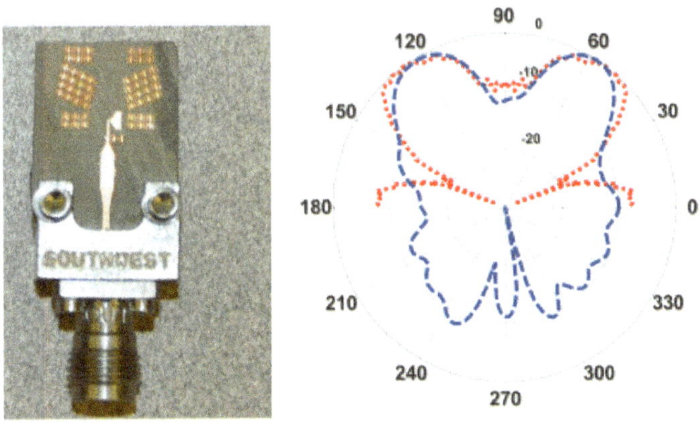


**Fig. 8.7** Dual-beam bow-tie antenna with artificially modified high permittivity loading on both sides and pattern at 60 GHz ([8], Reprinted with permission from IEEE)

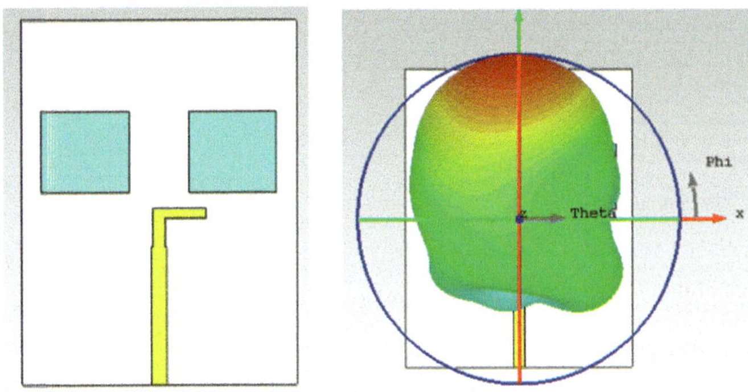


**Fig. 8.8** Dipole antenna loaded with the high-permittivity ($\epsilon_r = 7$) and the radiation pattern of the antenna at 28 GHz

The manipulation of electromagnetic waves using epsilon-near-zero (ENZ) media has been proposed and the reflection less propagation of the waves in highly compressed waveguide filled with ENZ is demonstrated in [12, 13]. The beam splitting using a hetero-junction of two anisotropic ENZ media with their principal axis rotated by some angle which is difficult to practically realize is reported in [14]. Also, Dadgarpour et al. [15] have used loading of ENZ metamaterials to generate dual beam radiation at 60 GHz by bending EM waves. However practically ENZ metamaterials have limited bandwidth and there is significant radiation along the broadside direction after 60 GHz which restricts the dual-beam bandwidth to 58–60 GHz only. In a similar way authors in [16] have used mu-near-zero loading to bowtie antenna to generate dual-beam radiation. Also, the design of ENZ media results in non-linearity from the simple relation between refractive index and epsilon

($n_y = \sqrt{\epsilon_y}$). Therefore, a small change in epsilon around zero results in large change in refractive index given as $\delta n = \frac{\delta A}{2n}$.

### 8.3.1 Dual-Beam Radiation Using Anisotropic Media

The technique of modulating phase velocities and separation of fields for dual-beam radiation using isotropic media on either side as presented above does not work at 28 GHz with a thin substrate. Therefore, an alternate technique for separating fields from a single dipole antenna for dual-beam radiation can be done using anisotropic media with high epsilon along the direction of polarization of the antenna. The theory and mechanism of dual-beam generation using the anisotropic media placed in front of the antenna is described below.

The guiding property of the infinitely anisotropic media is discussed in Chap. 2 for guiding of fields along the optical direction. The single epsilon high (Single-EH) anisotropic media (having one component of permittivity higher than other two) discussed in Chap. 2 can be used to guide the radiated fields from an antenna along optical direction of the media. In order to study the transmission and reflection of the waves incident on the single-EH media, we have used a Gaussian beam with in-plane electric field as source incident on the uniaxial medium as depicted Fig. 8.9a. The propagation of the wave from free space to the medium can be predicted by the Fresnel coefficients for the anisotropic medium. For TM polarized wave inside the medium, the effective refractive index N along the angle the refracted ray makes with the x-axis in the anisotropic media and is given as:

$$\mathrm{N} = \frac{\sqrt{\epsilon_x \epsilon_\perp}}{\sqrt{\epsilon_\perp \sin^2 \varphi + \epsilon_x \cos^2 \varphi}} \tag{8.1}$$

where $\epsilon_x$ and $\epsilon_\perp$ are the relative permittivity along optical axis (x-axis) of the medium and relative permittivity transverse to optical axis (y-axis), respectively. Figure 8.9b shows the comparison of the propagation of TM waves with in-plane electric field ($E_x$ and $E_y$) and transverse magnetic field ($H_z$) from air to isotropic media (first row) and to single-EH media (second row) with slanted boundary. We can see that the power flow in case of isotropic media is along y-direction, while as in case of single-EH media, power flow is more along x-direction due to the high epsilon value along this direction which results in the routing of fields along the x-axis. The power flow along x- and y-directions in case of single-EH media is titled more towards the optical axis. Therefore, the fields can be controlled by using high-epsilon media and propagation of EM waves can be routed along the optical axis. This routing of the EM waves would be used to generate dual-beam radiation by using two separated single-EH anisotropic medias loaded on either side of an EM source as shown in Fig. 8.10. Here we have used a Gaussian beam (waist $3\lambda_0$) with in-plane electric

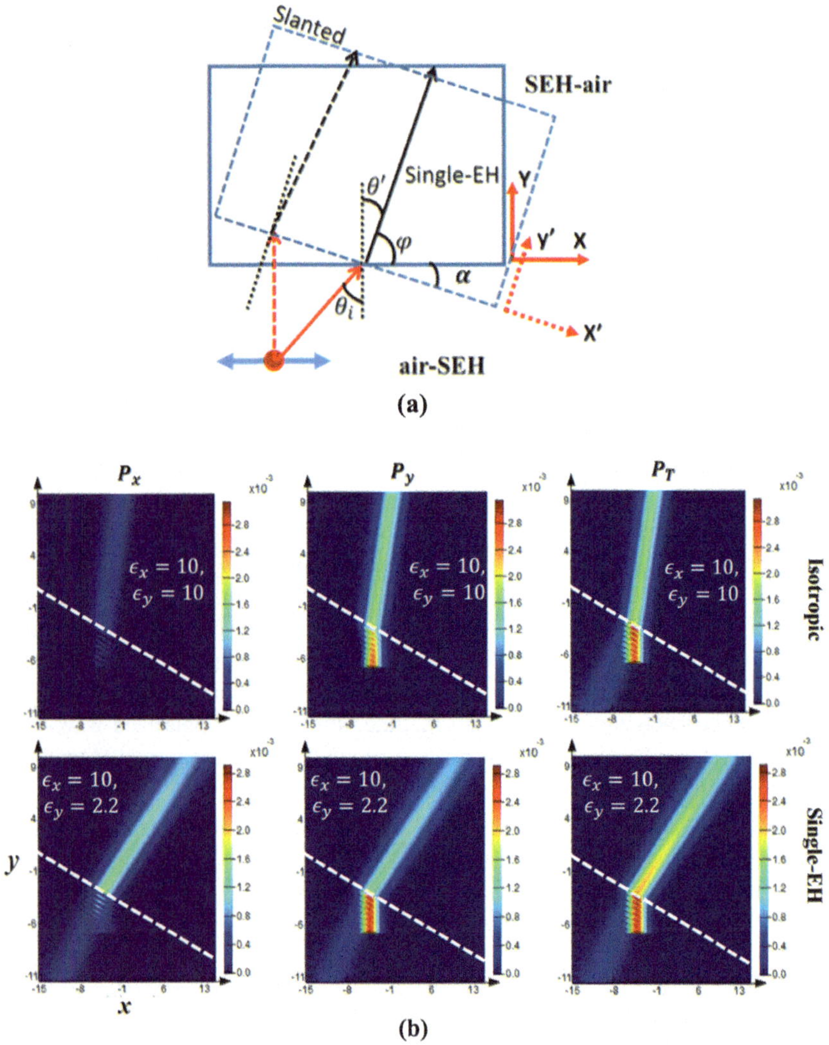


**Fig. 8.9** **a** Schematic of the plane waves excited on the single-EH media with and without slant and **b** comparison of the power flow ($P_x$—power along x-direction, $P_y$—power along y-direction and $P_T$—total power) at the slanted boundary for isotropic media and single-EH media. ([17], Reprinted with permission from IEEE)

field as a source and two single-EH slabs are placed at a negative slope of $\alpha = 20°$ with respect to optical axis.

After the coupling of TM waves from the source to the media due to the finite transverse permittivity, the wave would propagate in the single-EH medium diffraction-free and can exit the anisotropic medium reflection free for the incident angles up to 20° in the single-EH medium. This angular range can be increased up to 30° to

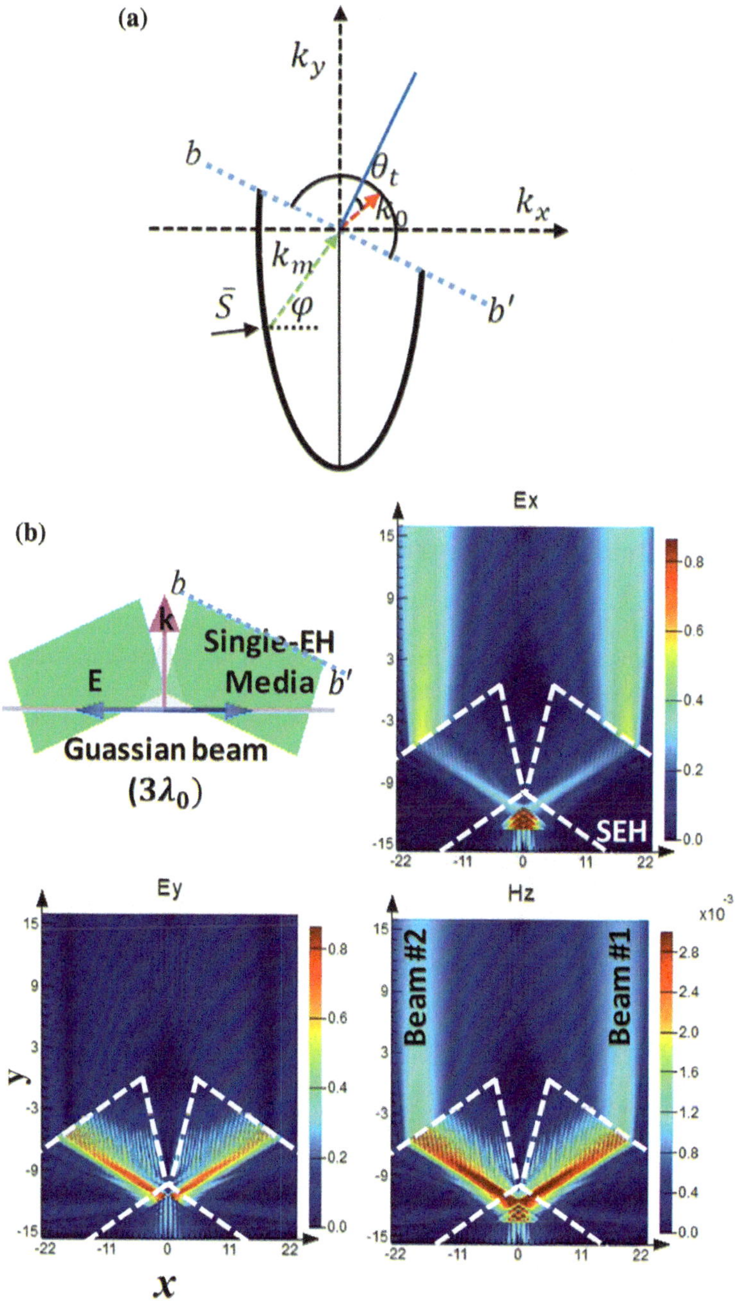


**Fig. 8.10** Dual-beam generation using single-EH media from a Gaussian beam **a** iso-frequency contour at exit boundary (bb′) between single-EH media (which is an ellipse) and free-space (a circle), **b** magnitude of electric field components and transverse magnetic field. ([17], Reprinted with permission from IEEE)

make most of the electric field in the single-EH media to couple to the free space by choosing a boundary at a negative slope to the optical axis of the single-EH medium.

The iso-frequency contour for TM waves at exit boundary is shown in the Fig. 8.10a, which depicts the effectiveness of choosing a boundary with negative slope $\alpha$. At the exit boundary, extraordinary waves exiting the single-EH medium, tangential phase components of waves in single-EH and the exit medium should be equal and gives the condition for transmission as:

$$\sin\theta_t = N_\varphi \cos(\varphi + \alpha) \tag{8.2}$$

where $\varphi$ is the angle which the wave vector in single-EH makes with the optical axis of the medium, $\alpha$ is the angle between the boundary and optical axis and $N_\varphi$ is the effective refractive index of single-EH medium along $\varphi(= 90 - (\alpha + |\theta_{rair\text{ - }SEH}|)$ which is given by (8.1).

## 8.4 Dual-Beam Dipole Antenna with Single-EH Media

The technique of dual beam generation using the anisotropic single-EH media from a Gaussian beam can be extended to antenna applications at mmWave frequencies. The process involves an end-fire antenna and a Single-EH media placed in front of the antenna with E-field along the optical axis of the media. Figure 8.11 shows the dipole antenna loaded with the anisotropic media with principal axis aligned with the global coordinate system of the antenna. To prove the concept discussed in Sect. 8.3, dipole antenna with single-EH media on either side of the dipole is designed in CST MWS with user defined anisotropic medium having relative permittivity $\epsilon_x = 10$ and $\epsilon_\perp = 2.2$ along the optical axis (x-axis) and transverse axis (y- and z-) respectively. Here we have used slanted single-EH media, by which most of the radiated fields

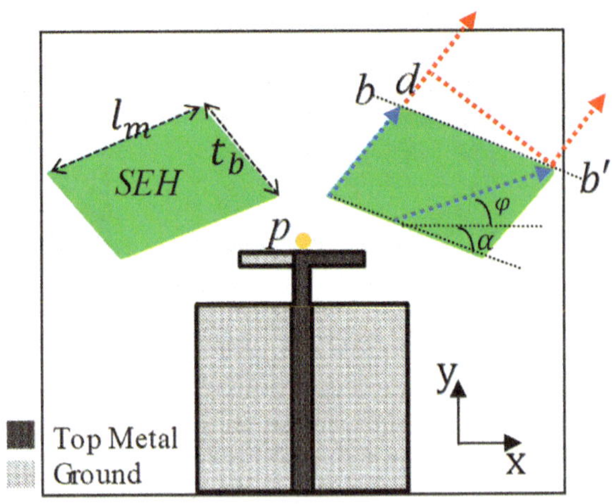


**Fig. 8.11** Geometry of the dipole antenna loaded with slanted single-EH media, ($l_m = 6.6$ mm, $t_b = 3.9$ mm, $\alpha = 20$) and point 'p' is the phase center of the antenna. ([17], Reprinted with permission from IEEE)

from the dipole antenna can coupled into the single-EH medium and can be routed along the optical axis to separate the radiated fields on the right and left.

The radiated fields coupled inside the Single-EH media and as predicted by theory of TM fields in the anisotropic media, in-plane Electric field and transverse magnetic field in an anisotropic media are given as [18]:

$$E(r) = Eo\frac{N}{\sqrt{n_x^2 + n_y^2 - N^2}}\left(y\frac{n_x}{n_y}\cos\varphi - x\frac{n_y}{n_x}\sin\varphi\right)e^{-ik.r} \tag{8.3}$$

$$H(r) = \frac{Eo}{\eta o}\frac{n_x n_y}{\sqrt{n_x^2 + n_y^2 - N^2}}ze^{-ik.r} \tag{8.4}$$

Here we can see that if $n_x = \sqrt{\epsilon_x}$ is more that the $n_y = \sqrt{\epsilon_\perp}$, Ex would be reduced by a factor $\frac{\epsilon_x}{\epsilon_\perp}$ and the power carried by the y-component of the electric field along the optic axis is reduced after entering the single-EH media as seen in Fig. 8.12. The graph shows the magnitude of the in-plane electric field components in the single-EH medium, most of the electric field is along y-axis resulting in the power flow along optic axis as seen from the figure.

This routing of the E-field along the transverse axis on either side splits the dipole antenna fields and results in the dual-beam radiation. The radiation pattern of the antenna with user-defined single-EH media is shown in Fig. 8.13. The radiation beams are directed along 55° and 125° in the azimuthal plane. These beam angles result from the phase difference of the magnetic field vectors on the exit boundary ($bb'$) after traveling in the single-EH medium. As we have seen the wave vectors with incident angle 0°–30° at the exit boundary would get transmitted. Thus, the maximum phase difference between the fields on $bb'$ with incident angle 0° and 30° can be obtained as:

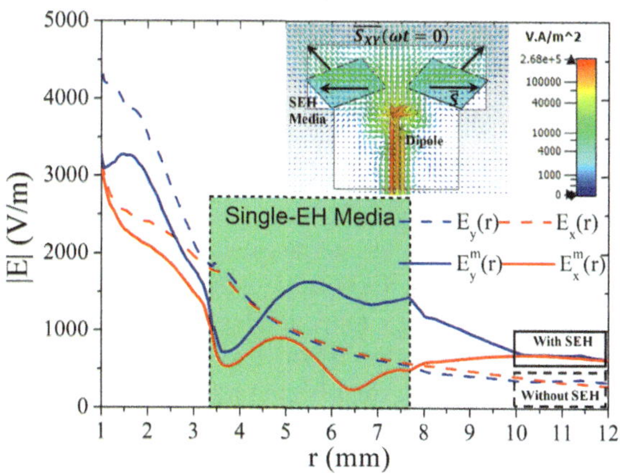


**Fig. 8.12** Magnitude of in-plane electric field components along the radial direction from the phase center with $\phi = 45°$ and $\theta = 90°$ along with the pointing vector in XY-plane. ([17], Reprinted with permission from IEEE)

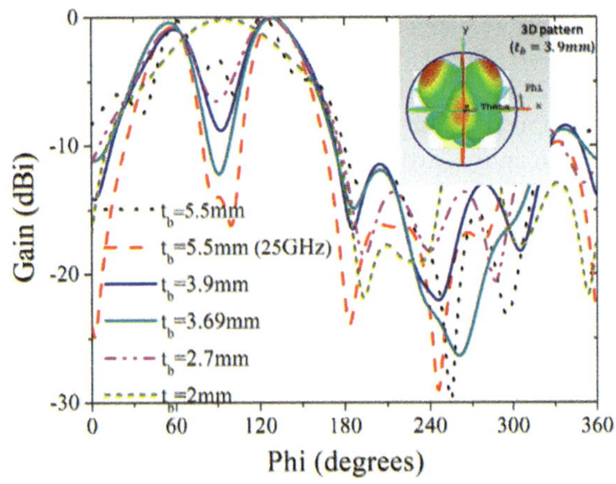


**Fig. 8.13** Normalized radiation pattern of the antenna with single-EH media along with the 3D radiation pattern at 28 GHz for different slab thickness ($t_b$). ([17], Reprinted with permission from IEEE)

$$\delta = k_0 t_b \left[ N_{\varphi=70} - \frac{N_{\varphi=40}}{\cos 30} \right] \tag{8.5}$$

where $t_b$ is thickness of the medium. The path difference between the waves after exiting the medium can be calculated from the phase difference in the single-EH medium as:

$$d = \frac{\lambda_0 \delta}{2\pi} \tag{8.6}$$

which gives the beam tilting angle as ($\phi = \cos^{-1} \frac{d}{l_m} - \alpha$). By substituting the values of N for $\varphi = 70°$ and $40°$ in (8.1) gives the phase difference at the exit boundary, $\delta = 68.43°$ from (8.5), which corresponds to the path difference of $d = \frac{\lambda_0}{5.26}$ from (8.6). The equal phase front from the waves exiting from the right single-EH medium is formed at 72.05° resulting in $\phi = 52.05°$. Similarly, the waves exiting from the left single-EH medium will form a beam along $\phi = 127.95°$. Hence two symmetrical beams are formed along $\pm$ 37.95° with respect to end-fire direction (y-axis).

Also, the effect of thickness of the single-EH media on the radiation pattern can be seen from Fig. 8.13, as the thickness is increased above 2 mm dual-beam radiation pattern at 28 GHz is formed and the pattern deteriorates above 5.5 mm; however dual-beam radiation is formed at this thickness for 25 GHz. This is due the fact that reflection at multiple interfaces can be made zero when $N_\varphi t_b = m \frac{\lambda_0}{2}$ is satisfied. Equivalently when the optical length of the single-EH media equals half of the free-space wavelength $\lambda_0$, reflection less transmission is achieved, and dual-beam can be formed. Here $N_\varphi$ is the effective refractive index and varies between 1.8 and 2.6 and $m$ is an integer. For dual-beam radiation pattern at 28 GHz, $t_b$ should be more than 2 mm (for $m = 1$). We have chosen the thickness of the medium 3.9 mm so that 26 to 30 GHz dual-beam-bandwidth is achieved.

## 8.5 Practical Implementation and Dual-Beam Antenna

The technique and simulation results presented in previous section can be implemented practically using a substrate integrated waveguide (SIW) dipole antenna and Single-EH media. Here the Single-EH media used in simulations can be realized using metamaterial slabs with anisotropic unit cells. The complete realization and design of Single-EH media using metamaterials/dielectric slabs is discussed in Chap. 2. The results with the metamaterials ELC slabs loaded to an SIW dipole antenna are discussed below.

The dipole antenna is loaded with user-defined material in the simulation software and the results obtained agree well with the proposed theory. We have seen that the transverse magnetic waves effectively couple to the single-EH media and are routed along the optical axis to separate the fields. For practical purpose we have chosen SIW dipole antenna as a TM source to generate dual-beam radiation with lower front-to-back ratio and good bandwidth at 28-GHz frequency band.

Also, SIW dipole antenna configuration at mmWave frequencies did not suffer with the radiation loss from the feed as it is observed that the unwanted radiation from the dipole antenna-feed used in simulation above affects the radiation pattern of the antenna along broadside direction. Therefore, SIW dipole antenna is a better choice and is designed using standard formulas on NY9220 substrate with height of 0.254 mm and having relative permittivity 2.2 as shown in Fig. 8.14 [19]. A microstrip transition is used to feed the dipole antenna. The measured and simulated $|S_{11}|$ of the SIW antenna is plotted in Fig. 8.15, operating bandwidth of 24.5 to 31 GHz is obtained with this configuration. The E-plane radiation pattern of the SIW antenna is plotted in Fig. 8.16 (No slab case), we can see that most of the energy ($\sim 80\%$) from the dipole is radiated in the angular range, $\theta_i = \pm 50°$ from y-axis (that is dipole radiates 80% of the power along $\phi = 40°$ to $140°$).

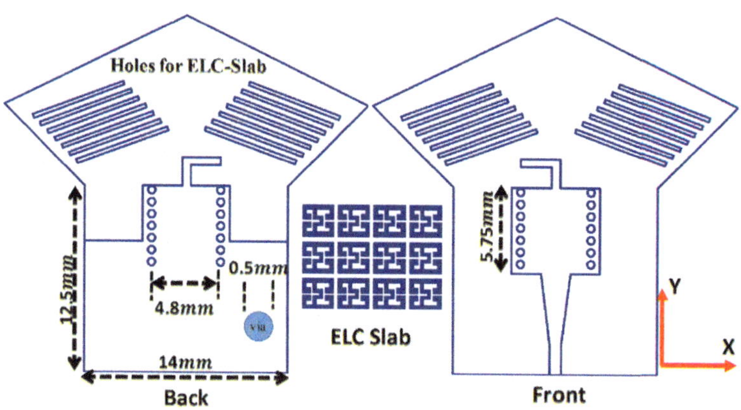


**Fig. 8.14** Geometry of the proposed SIW dipole antenna loaded with 7-metamaterial ELC-slabs. ([17], Reprinted with permission from IEEE)

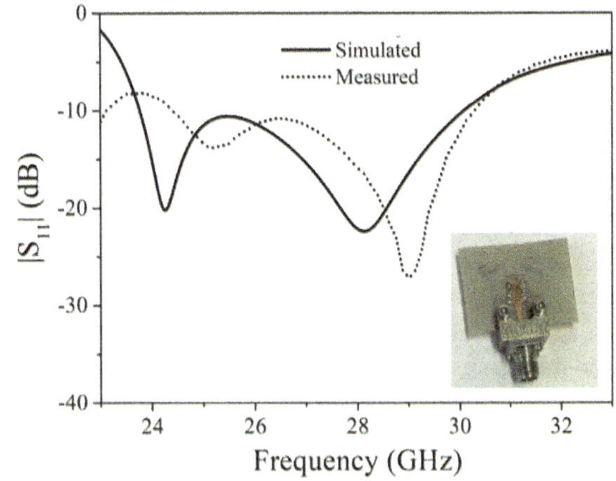


**Fig. 8.15** Measured and simulated reflection coefficient of SIW dipole antenna. ([17], Reprinted with permission from IEEE)

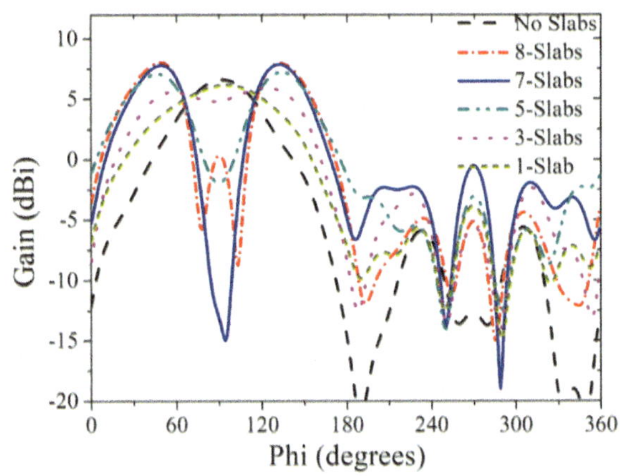


**Fig. 8.16** E-plane radiation pattern of SIW dipole antenna loaded with ELC slabs at 28 GHz. ([17], Reprinted with permission from IEEE)

The antenna is then loaded with the array of metamaterial ELC slabs on either side to rout the fields along the media. The E-plane radiation pattern of the antenna loaded with the vertical ELC slabs is shown in Fig. 8.16. The trend in the plots show that with increasing/decreasing the ELC slabs, dual-beam radiation at lower frequencies (26 GHz)/higher frequencies (29 GHz) can be generated. Loading with 7-slabs (corresponds to single-EH medium thickness of $t_b = 4$ mm), dual-beam radiation is generated for 27–30 GHz, which is in accordance with the results when dipole was loaded with user-defined anisotropic media and the theory of reflection less transmission at multiple interfaces. We have seen by reducing the ELC slab to 5, dual-beam at higher frequencies above 30 GHz can be generated however at lower frequencies around 26 GHz the thickness of the medium becomes smaller than the optical length at 27 GHz, dual-beam is not generated.

The E-field and pointing vector distribution at a cut plane $z = 0.254$ mm at $\omega t = \pi/4$ is shown in Fig. 8.17, the electric field in the single-EH media is seen to

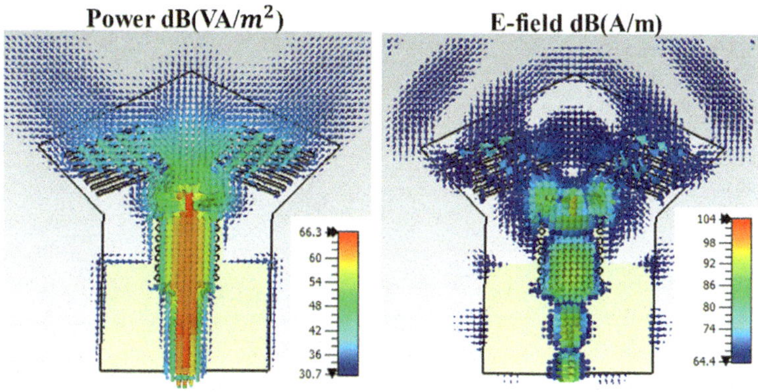


**Fig. 8.17** Poynting vector and electric field in XY plane at $\omega t = \pi/4$

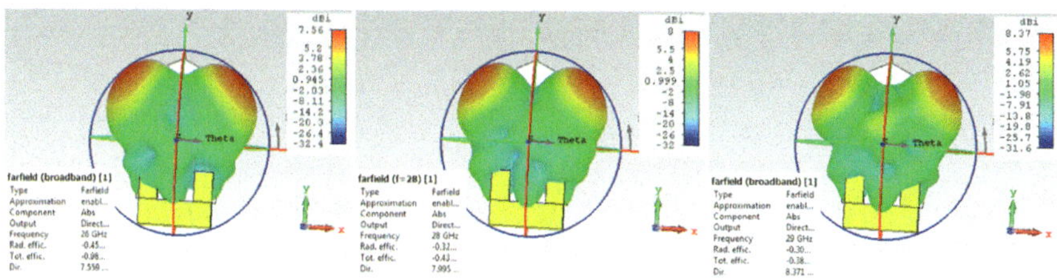

**Fig. 8.18.** 3D radiation pattern at 26, 28 and 29 GHz with 7-ELC slabs loaded to the SIW dipole antenna. ([17], Reprinted with permission from IEEE)

be oriented mostly along y-axis and H-field normal to the plane which results in the power flow in x-direction which is the optic axis of the medium. Thus, the radiated energy is guided along the two SEH media resulting in the dual-beam radiation.

The 3D radiation pattern of the antenna is shown in Fig. 8.18, dual-beam radiation is generated for 26–30 GHz with a beam peak of about 8 dBi when seven ELC slabs are incorporated with the antenna and has radiation efficiency (simulated) above 90% in the band. As the dual-beam radiation bandwidth depends on the thickness of the medium, we can achieve wide dual-beam radiation for 26–31 GHz by just increasing or decreasing the number of ELC slabs.

The configuration of the SIW dipole antenna employing single-EH anisotropic media realized using the vertical metamaterial ELC-slabs is shown in Fig. 8.19a. The ELC slabs are supported using the holes made in the antenna substrate and were glued to the antenna substrate as it was very difficult to maintain a gap of 0.254 mm between the slabs with this thin substrate. First the SIW dipole antenna was designed and characterized using a 2.92 mm connector and then ELC-slabs are loaded to antenna.

The measured port performance of the antenna with seven ELC-slabs is presented in the Fig. 8.19b. The loading of ELC-slabs does affect the reflection coefficient around 24 GHz of the antenna and this effect can be minimized by using single-EH

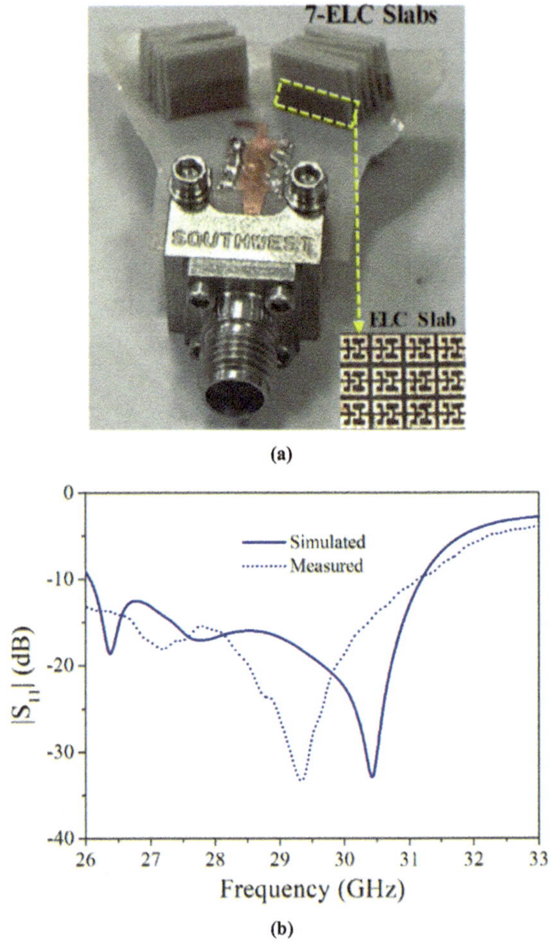


**Fig. 8.19** **a** Photograph of SIW dipole antenna with seven ELC slabs, **b** simulated and measured reflection coefficient of the SIW dipole antenna loaded with seven ELC-slabs. ([17], Reprinted with permission from IEEE)

media with lower transverse relative permittivity however, 26–31 GHzimpedance bandwidth is achieved with return loss > 10 dB. The prototype antenna is tested in a far-field type anechoic chamber with a separation of about 80 cm between the standard Ka-band horn antenna and proposed antenna.

The radiation pattern for 27, 28, 29 and 30 GHz frequency samples is plotted with the simulated results. The dual-beam radiation pattern is observed for the band 27–30 GHz with magnitude along end-fire direction less than − 10 dB as shown in Fig. 8.20. The proposed antenna has dual-beam radiation characteristics with main beam peaks directed at $\phi = 50°$ and 130°.

After that, one ELC slab is added on each side and radiation performance is again measured over 26–28.5 GHz, this time with eight ELC slabs on both sides. The measured and simulated radiation patterns shown in Fig. 8.21 verify the dual-beam pattern for 27–28.5 GHz; however, at 29 GHz some part of the fields gets leaked

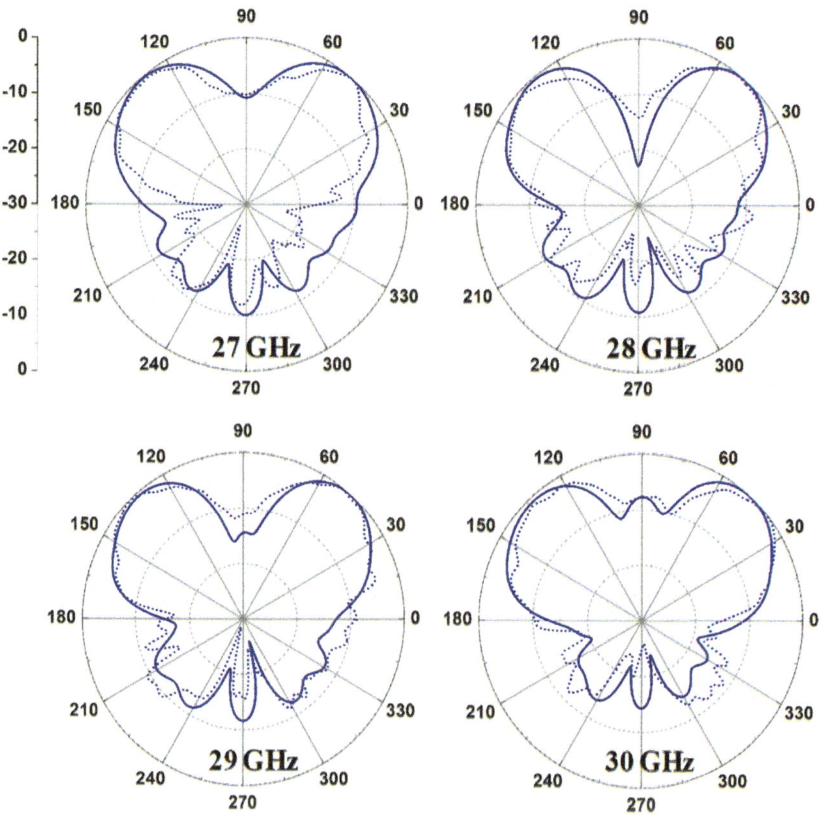


**Fig. 8.20** Measured and simulated radiation pattern (E-plane) of the SIW dipole antenna with 7-ELC metamaterial slabs at 27, 28, 29 and 30 GHz. (Solid: simulated; dashed: measured). ([17], Reprinted with permission from IEEE)

through the metamaterial loading as the single-EH media thickness is more than the optical length at 29 GHz.

The measured gain of the antenna with 7-ELC slabs is shown in Fig. 8.22, measured realized gain of 7.93 dBi is achieved at 28.5 GHz without any reflectors at the back of the dipole antenna. With this technique of loading with single-EH media, the dual-beam radiation bandwidth can be shifted to cover lower or higher frequencies by just increasing or decreasing the number of ELC-slabs, this way we can cover 26–31 GHzbandwidth.

In the last part of this chapter, the technique of dual-beam generation using uniaxial anisotropic high epsilon media is compared with the techniques discussed in Sect. 8.3 such as isotropic dielectric loading on either sides and use of ENZ/MNZ media over an antenna. Some of the features and requirements for these techniques are given in the following Table 8.1. The Single-EH media loading technique does offer some advantages like large impedance BW and dual-beam BW (with radiation null < –

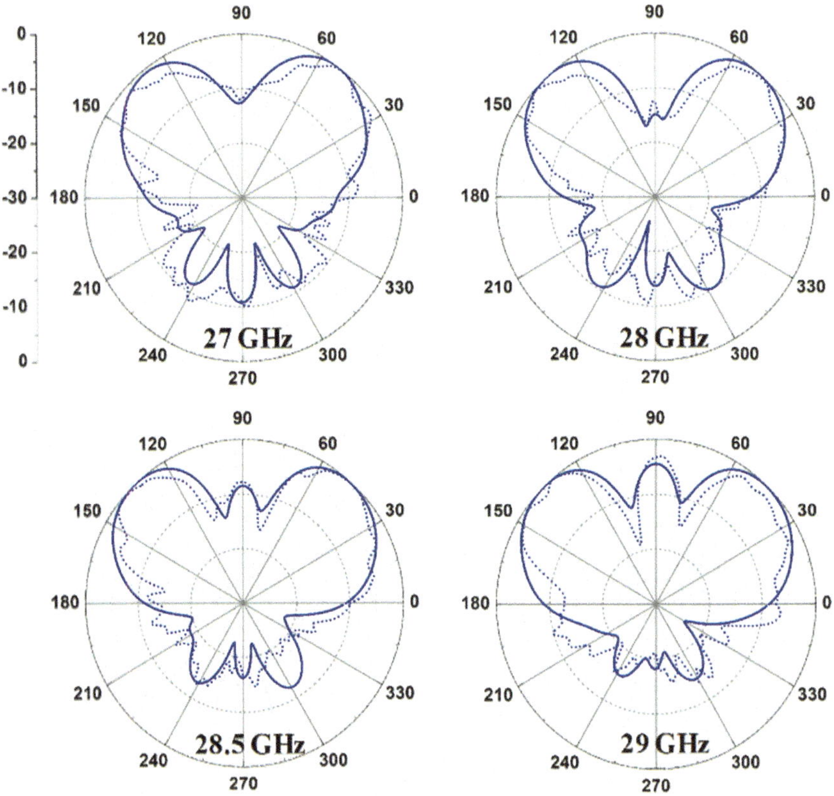


**Fig. 8.21** Measured and simulated radiation pattern (E-plane) of the SIW dipole antenna with 8-ELC metamaterial slabs at 27, 28, 28.5 and 29 GHz. (Solid: simulated; dashed: measured). ([17], Reprinted with permission from IEEE)

10 dB). Moreover, the dual-beam BW can be tuned to lower or higher frequencies by adjusting the number of slabs or media thickness.

## 8.6 Conclusion

This chapter presents dual-beam antenna for mmWave applications. Various techniques of dual beam generation are presented. First a simple technique using two dipole antenna arrays to generate dual beam radiation is discussed. Other techniques using metamaterials to separate the fields from the single end-fire antenna to generate dual-beam radiation are discussed along with their shortcomings. Lastly a technique based on uniaxial anisotropic media for dual-beam radiation is discussed in detail with measured results. The implementation of the technique using metamaterials

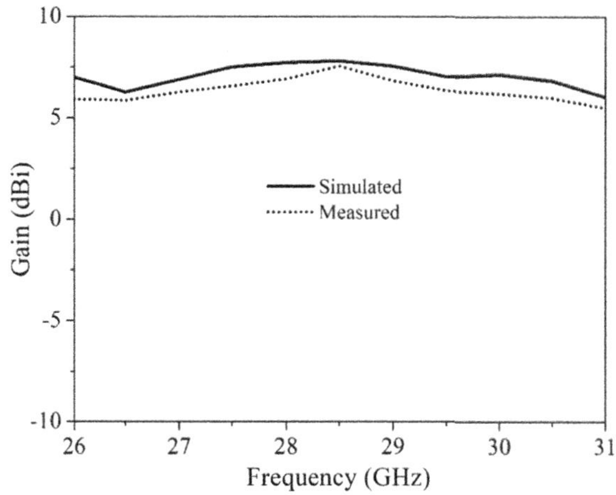


**Fig. 8.22** Measured and simulated gain of the antenna with seven ELC slabs loaded to SIW dipole antenna. ([17], Reprinted with permission from IEEE)

**Table 8.1** Comparison of different techniques for dual-beam generation

| Technique used | Design | Requirements | Realization | Radiation pattern |
|---|---|---|---|---|
| High permittivity for TE surface waves (Ref. [8]) | Simple | Employs TE surface waves, dielectric substrate with high epsilon and height ($h \gg \frac{\lambda_0}{10}$) | Here metamaterial unit cells are employed to control TE waves on both sides | Dual-beam with significant radiation along the end-fire direction at low frequencies |
| ENZ media loaded to slot (Ref. [15]) | Complex | Fields couple to ENZ media, but media loading will affect the antenna return loss as ENZ has large impedance. Also, BW of ENZ is limited | Meander-line unit cell with ENZ is used, but bulk properties of the media is not discussed | Dual-beam rad with significant radiation along the broadside. Limited DB-radiation BW |
| Single-EH anisotropic media loaded to dipole to route the fields along the optical axis | Moderate | Fields are coupled to the media easily by choosing optic axis along the E-field | Asymmetric ELC unit cells are used to mimic exact media used in simulations | Dual-beam radiation with variable radiation bandwidth by adjusting media thickness |

already discussed in Chap. 2 is employed to generate dual-beam radiation in the band 27–30 GHz.

## References


1. Li K-H, Ingram MA, Rausch EO (2002) Multibeam antennas for indoor wireless communications. IEEE Trans Commun 50(2):192–194. https://doi.org/10.1109/26.983314
2. Kirthiga S, Jayakumar M (2014) Performance of dualbeam MIMO for millimeter wave indoor communication systems. Wirel Commun 77(1):289–307. https://doi.org/10.1007/s11277-013-1506-0
3. Ayach OE, Rajagopal S, Abu-Surra S, Pi Z, Heath RW (2014) Spatially sparse precoding in millimeter wave MIMO systems. IEEE Trans Wirel Commun 13(3):1499–1513. https://doi.org/10.1109/TWC.2014.011714.130846
4. Hussain R, Alreshaid AT, Podilchak SK, Sharawi MS (2017) Compact 4G MIMO antenna integrated with a 5G array for current and future mobile handsets. Antennas Propag IET Microw 11(2):271–279. https://doi.org/10.1049/iet-map.2016.0738
5. Alreshaid AT, Hussain R, Podilchak SK, Sharawi MS (2016) A dual-element MIMO antenna system with a mm-wave antenna array. In: 2016 10th European conference on antennas and propagation (EuCAP), pp 1–4. https://doi.org/10.1109/EuCAP.2016.7481146
6. Sun S, Rappaport TS (2013) Multi-beam antenna combining for 28 GHz cellular link improvement in urban environments. In: 2013 IEEE global communications conference (GLOBECOM), pp 3754–3759. https://doi.org/10.1109/GLOCOM.2013.6831657
7. Wani Z, Abegaonkar MP, Koul SK (2018) Dual-beam antenna for millimeter wave MIMO applications. In: 2018 IEEE indian conference on antennas and propogation (InCAP), pp 1–3. https://doi.org/10.1109/INCAP.2018.8770883
8. Dadgarpour A, Zarghooni B, Virdee BS, Denidni TA (2016) Single end-fire antenna for dual-beam and broad beamwidth operation at 60 GHz by artificially modifying the permittivity of the antenna substrate. IEEE Trans Antennas Propag 64(9):4068–4073. https://doi.org/10.1109/TAP.2016.2574881
9. Hong W et al (2017) Multi-beam antenna technologies for 5G wireless communications. IEEE Trans Antennas Propag PP(99):1–1. https://doi.org/10.1109/TAP.2017.2712819
10. Liu P, Zhang Z, Li Y, Feng Z (2017) A dual-beam eight-element antenna array with compact CPWG crossover structure. IEEE Antennas Wirel Propag Lett 16:1269–1272. https://doi.org/10.1109/LAWP.2016.2632123
11. Alexopoulos NG, Katehi PB, Rutledge DB (1983) Substrate optimization for integrated circuit antennas. IEEE Trans Microw Theory Tech 31(7):550–557. https://doi.org/10.1109/TMTT.1983.1131544
12. Edwards B, Alù A, Young ME, Silveirinha M, Engheta N (2008) Experimental verification of epsilon-near-zero metamaterial coupling and energy squeezing using a microwave waveguide. Phys Rev Lett 100(3):033903. https://doi.org/10.1103/PhysRevLett.100.033903
13. Silveirinha M, Engheta N (2006) Tunneling of electromagnetic energy through subwavelength channels and bends using ε-near-zero materials. Phys Rev Lett 97(15):157403. https://doi.org/10.1103/PhysRevLett.97.157403
14. Memarian M, Eleftheriades GV (2015) Analysis of anisotropic epsilon-near-zero hetero-junction lens for concentration and beam splitting. Opt Lett 40(6):1010–1013. https://doi.org/10.1364/OL.40.001010
15. Dadgarpour A, Sorkherizi MS, Denidni TA, Kishk AA (2017) Passive beam switching and dual-beam radiation slot antenna loaded with ENZ medium and excited through ridge gap waveguide at millimeter-waves. IEEE Trans Antennas Propag 65(1):92–102. https://doi.org/10.1109/TAP.2016.2631140

16. Dadgarpour A, Sorkherizi MS, Kishk AA, Denidni TA (2016) Single-element antenna loaded with artificial mu-near-zero structure for 60 GHz MIMO applications. IEEE Trans Antennas Propag 64(12):5012–5019. https://doi.org/10.1109/TAP.2016.2618485
17. Wani Z, Abegaonkar MP, Koul SK (2020) Dual-beam antenna using routing of electromagnetic waves by single-epsilon-high anisotropic medium at 28 GHz. IEEE Trans Antennas Propag 68(1):142–151. https://doi.org/10.1109/TAP.2019.2938793
18. Kong JA (1986) Electromagnetic wave theory. Wiley [Online]. Available: https://books.google.co.in/books?id=OposAAAAYAAJ
19. Djerafi T, Doghri A, Wu K (2014) Substrate integrated waveguide antennas. In: Chen ZN (ed) Handbook of antenna technologies. Springer, Singapore, pp 1–60. https://doi.org/10.1007/978-981-4560-75-7_57-1

# Chapter 9
# Millimeter Wave Lens Antennas

## 9.1 Introduction

Millimetre waves can provide exploding capacity demands for next generation wireless systems. Also due to the short wavelengths, full dimensional massive MIMO operation becomes possible as large number of antennas can be packed in a small form factor. The MIMO techniques discussed in previous chapter can be efficiently applied to maintain the link between the transmitter and the receiver with better single-to-noise ratio [1, 2]. To exploit the full potential of MIMO technique, MIMO is usually implemented digitally with one RF chain for each transmitter/receiver antenna. However, implementing one RF chain for each antenna would increase cost tremendously and the power consumption would deteriorate the SNR significantly at mmWaves. To reduce the hardware cost and power consumption, analog beamforming presents a suitable array gain for better SNR using phase shifters and one RF chain. The signal processing complexity and hardware cost can be reduced significantly by employing lens antenna array for large MIMO systems. The electromagnetic lens having larger aperture provides array gain replacing phase shifters at mmWaves. The spatial multiplexing which cannot be achieved using analog beamforming can be implemented using lens antenna arrays with fewer RF chains feeding the antennas.

The lens antenna arrays consist of EM lens with an array of antenna elements in the focal plane of the lens. The lens play an important role to steer the radiation beam to different direction depending on which antenna element is excited to achieve angle dependent radiation beams. The EM lens with the antenna array can be implemented using conventional dielectric materials with thickness profile, however the resulting lens are bulky and non-planar. This chapter presents the technique of lens antenna array for mmWave MIMO and design of planar metasurface lens which replaces the bulky dielectric lenses.



S. K. Koul and Z. Wani, *Novel Millimetre Wave Antennas for MIMO and 5G Applications*, Lecture Notes in Electrical Engineering 819,
https://doi.org/10.1007/978-981-16-7278-1_9

## 9.2 Lens Antennas for mmWave MIMO

Massive MIMO consists of large number of antennas at the base station for array gain and beamforming. Through this we can achieve desirable performance at mmWave frequencies with good SNR and can service multiple users at a time. As the mmWave links are power limited than bandwidth limited, thus achieving an array gain is very important for adequate SNR. The beamforming can be done is three ways analog, digital and hybrid with each having their own advantages. The digital beamforming technique requires one dedicated RF chain for TX/RX antenna thus is not feasible for massive MIMO due to huge cost and power consumption. Analog beamforming uses one RF chain for array gain at Tx/Rx, however the performance is limited by the efficient phase shifters at mmWave frequencies. The advantages of both these techniques can be achieved using hybrid beamforming where fever number of the RF chains are used, and array gain is achieved due to analog beamforming using phase shifters [3].

The lens antenna array based on concept of beamspace MIMO efficiently gives access to p communication modes of an n-dimensional mmWave link ($n > p$) as shown in Fig. 9.1. These p modes are then mapped to the n feed signals on the focal surface of the lens. The beamspace MIMO uses hybrid analog–digital front end at the transmitter in which a lens is used for analog spatial beamforming to transform signals from antenna space to the beamspace [4, 5]. In case of beamspace MIMO, the beamforming in a particular direction (AoA/AoD) is done by the lens (which provides power gain for power limited mmWave links) depending on which antenna is selected; this significantly reduces the RF chain cost. While as in the case of conventional beamforming techniques, array gain is achieved using simultaneously feeding number of sub array antennas with true time delay devices (for wide bandwidths).

The capacity performance of a beamspace MIMO is compared with the conventional MIMO antenna systems in [4]. For a conventional MIMO antenna system with $p_{max}$ antennas and the maximum number of digital modes is $p_{max}$. The capacity of conventional MIMO system is given by $C = p_{max} \log(1 + \rho)$, where $\rho$ is SNR.

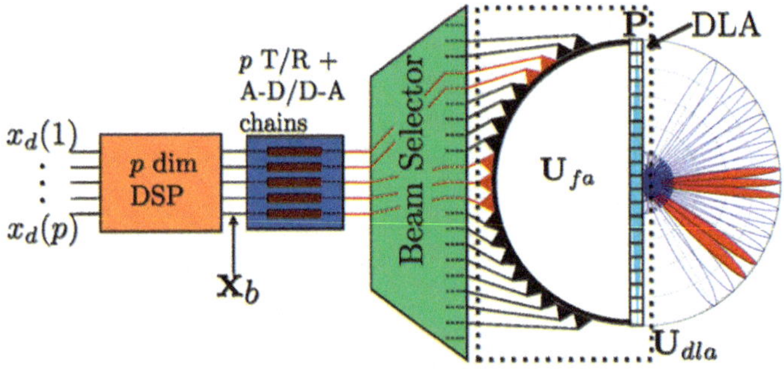


**Fig. 9.1** Beamspace MIMO configuration ([5], Reprinted with permission from IEEE)

The beamspace MIMO uses the lens for archiving higher power gain with those of conventional MIMO. The number of data streams here is p ($< 1 < p < p_{max}$). The capacity of beamspace MIMO is given by: $C = p\log\left(1 + \rho n^2/\text{ppmax}\right)$. In case of beamspace MIMO, the SNR term reflects SNR enhancement than the conventional MIMO with isotropic antennas. The comparison of the capacities between the conventional MIMO and beamspace MIMO is presented in references [4, 5] for different SNR values. It has been shown that the capacity of the beamspace MIMO (with $p = p_{max}$) exceeds the performance of the conventional MIMO systems.

The signal processing complexity and RF chain cost can be reduced without considerable loss in performance using lens MIMO antenna array. The array gain can be easily obtained by use of the lens which is basically a variable phase shifter for EM rays for different directions. The phase shifters in case of massive MIMO can be replaced by a passive lens which significantly would reduce the cost [6]. The new path division multiplexing (PDM) scheme in [6] for mmWaves has better ISI rejection and low RF chain cost. This multiplexing technique requires the design of antenna arrays with EM lens to focus different multipath signals to different antennas based on the placement of the antenna which also results in wide-scan radiation characteristics.

## 9.3 Planar Metasurface Lens Design

The EM lenses play a crucial role in mmWave lens MIMO to manipulate the radiated energy from antennas. The lens can be designed by utilizing the material properties and geometry of the lens to focus the energy to a particular direction. The mmWave frequency lenses are quite essential for 5G systems and realization using dielectric material and curved geometries are not preferred due to their bulky size. Also, the dielectric lenses are not matched which results in reflections at air-dielectric interface.

Metasurfaces are 2D arrays of sub-wavelength resonators which are very tiny and can control the transmission of the electromagnetic waves through them. These small scatters or unit cells manipulate the EM waves by introducing arbitrary reflections, polarization change, wavefront control at very compact profile. A planar EM metasurface lens can be designed using array of these small resonators for mmWave applications. Depending on the unit cell properties, the planar metasurface lens can be classified in to two groups:

### *9.3.1 Graded Refractive Index (GRIN) Lens*

The schematic of a graded refractive index lens shown in Fig. 9.2, consists of the array of unit cells whose effective refractive index can be configured to an any value in a range of values. The positive refractive index of such unit cell can be achieved by operating in the non-resonant region of the cell. The unit cell with metallic structure is chosen such that to have electric resonance at higher frequency than the operating

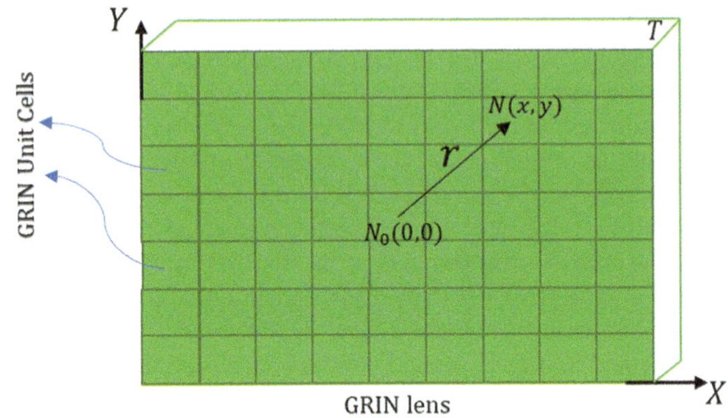


**Fig. 9.2** Schematic of the flat GRIN lens

range, this results in a higher refractive index for lower frequencies. A GRIN lens with flat profile for X-band applications is demonstrated in [7], the metamaterial unit cells consist of three rectangular loops on top and bottom of the substrate. The metamaterial unit cell shown in Fig. 9.3a has a periodicity of 4 mm and the resonant frequency can be tuned by the length 'a' of the loops. The extracted refractive index of the unit cell for different values of the length of the rectangular loops is shown in Fig. 9.3b. The refractive index variation of 1.2–3.7 is achieved by changing the length 'a' from 0.1 to 3.85 mm.

Now to design a planar lens using the array of such unit cells, the phase compensation is done using graded refractive index unit cells operating in non-resonant frequency region. Each unit cells refractive index is tuned using the length of the three rectangular loops on each side. The refractive index profile of the lens for a focal length $f$ is given by:

$$N(x, y) = N_0 - \frac{\sqrt{r^2 + f^2} - f}{T} \tag{9.1}$$

where T is the thickness of the lens and $N_0$ is the maximum value of the refractive index achieved with variation of the loop length. Here $r = \sqrt{x^2 + y^2}$, radius from the centre of the lens as shown in Fig. 9.2. The planar lens consists of 36 × 36 unit cells with each cell providing different refractive index from a range of 1.2 to 3.7 at each location given by (9.1) and the refractive index profile over the lens with a thickness of 12 mm as shown in Fig. 9.4. The GRIN lens is designed for a focal length of 100 mm and is loaded to a tapered slot end-fire antenna. By placing the lens in front of an antenna results in a peak gain of 21.2 dB at 9.5 GHz for an aperture size of 14.4 cm × 14.4 cm, resulting in an aperture efficiency of 50%. The higher aperture efficiency results due to the smaller unit cell size as they are tuned to be resonant at higher frequency range than the operating range. However, GRIN planar lenses usually have larger focal length and for a fixed aperture and fixed refractive

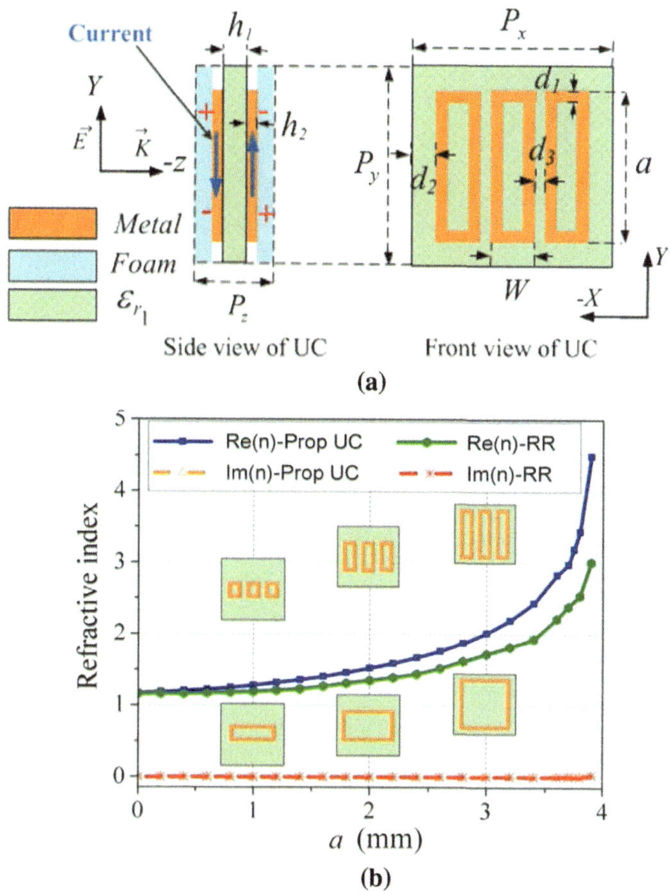


**Fig. 9.3** **a** GRIN unit cell with rectangular loops ($Px = Py = 4$ mm) and **b** extracted refractive index versus the loop length a ([7], Reprinted with permission from IEEE)

index variation, we can calculate the minimum focal length for the above lens. The minimum focal length can be given as:

$$f = \frac{r_{\max}^2}{2\Delta nT} - \frac{\Delta nT}{2} \tag{9.2}$$

where $\Delta n = N_0 - N(X, Y)$ is the difference in the refractive index in the centre and outer edge of the lens for a radius of $r_{\max} = \sqrt{X^2 + Y^2}$. For an example, let us now consider designing a GRIN lens having $f = 12$ mm at 28 GHz frequency with a thickness $T = 1.4$ mm and for the same $r_{\max} = 72$ mm. The refractive index variation (− 45 to 3.7) is required to implement the same profile as given by (9.2). This huge variation of the refractive index is impossible to realize using metamaterials. Therefore, the direct implementation of the phase profile using phase

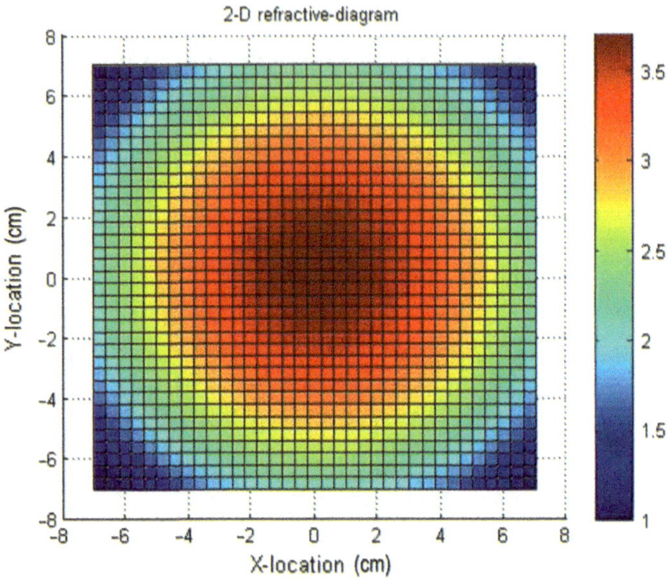


Fig. 9.4 Refractive index profile over the surface of the lens ([7], Reprinted with permission from IEEE)

tunned unit cell (PGMS) is easier than GRIN unit cells. The refractive index variation and phase variation for the two cases with different thickness and focal length is shown in the Fig. 9.5. With the setting of focal length, f = 12 mm and lens thickness (T = 1.4 mm), the required refractive index variation is huge which is practically not possible however, the phase variation can be achieved using the phase tuned unit cells by translating the phase required to (0°–360°) by adding $2n\pi$ to the phase.

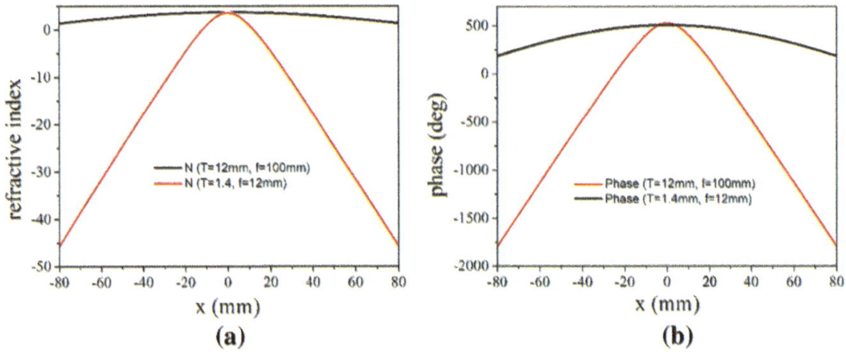


Fig. 9.5 Refractive index and phase variation over the lens for different focal length and thickness **a** GRIN lens and **b** phase tuned lens

### 9.3.2 *Phase Gradient Metasurface Lens*

This is most versatile technique to have planar MS lens using a phase tuned unit cell. The phase gradient metasurface (PGMS) lens consists of two-dimensional array of phase tuned unit cells which control the transmission phase of propagating wave at each unit cell location Theoretically any f/D lens can be designed using this technique however one of the main drawbacks is the unit cells are of the size of $\lambda/2$ due to operation in the resonant region.

The configuration of the planar lens on top of the antenna array is shown in Fig. 9.6. The configuration shown consists of planar lens placed above the antenna array at a focal length of $f$. The three antennas placed in the focal region of the lens would receive incoming wave from three different directions or can transmit energy along different directions resulting due to the phase compensation due to the lens. Conventionally the phase correction in case of dielectric lens is done using the curvature of the lens however metasurface (MS) consisting of array of subwavelength resonators providing phase delays to transform spherical waves into planar waves resulting in an ultra-thin planar lens configuration. By using planar MS, the phase of the outgoing waves can be locally controlled at each point on the surface using the geometry of the unit cells. Figure 9.6 shows the planar lens for transforming the spherical waves from the antennas into the plane waves after passing through the lens. The phase compensation can be done using the phase discontinuities introduced by the small MS unit cells. The planar MS lens is designed by arraying the unit cells along x- and y-direction as shown in top view of the lens. The phase discontinuity ($\phi(x, y)$) provided by each unit cell can be calculated using the Fermat's principle to compensate for different path lengths as:

$$\phi(x, y) = -\frac{2\pi}{\lambda}\left(\sqrt{r^2 + f^2} - f\right) \pm 2n\pi \tag{9.3}$$

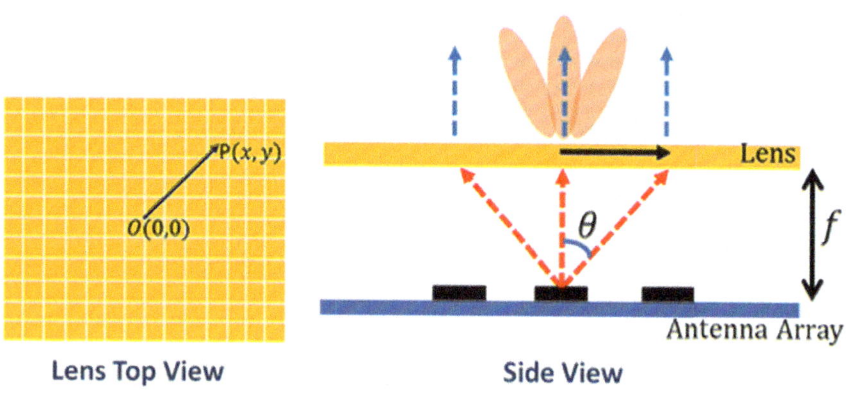


**Fig. 9.6** Configuration of the PGMS lens. ([8], Reprinted with permission from IEEE)

$$r = \sqrt{x^2 + y^2} = f \tan\theta \tag{9.4}$$

where $f$ is the focal length of the lens and $r$ is the distance between center of the lens and point $P(x, y)$. Using (9.3), the phase can be calculated for each unit cell in the lens for the phase compensation. The phase delay profile is directly implemented using the phase tuned unit cells to cover entire 360 phase with no transmission losses and perfect matching. The transmission losses through the unit cells results in loss in the transmitted power, therefore it is required that the unit cells have lower reflection coefficient and transmission coefficient close to 1. Also, other important parameter while designing a unit cell for PGMS lens is the size of the unit cells which decides the phase quantization errors to implement the phase profile smoothly. The phase errors resulting due to the larger size unit cells affects the aperture efficiency of the lens. The design of phase gradient unit cells to achieve transmission phase delays based on $S_{21}$ phase, requires having a wide transmission band with reconfigurable bandwidth. The transmission band is shifted to higher or lower frequencies around the operating band by the changing the length of the unit cell. The PGMS unit cells operate in the resonant region to have variable phase delays and thus are larger in size than GRIN unit cells and thus the PGMS lenses have smaller aperture efficiencies.

## 9.4 Ultra-Thin Planar Metasurface Lens

The planar ultra-thin lens is designed for focal length of 12 mm and consists of $15 \times 15$ metasurface (MS) unit cells. The designed unit cell is basically a centre connected N-pole structure consisting of two crossed dipoles with curved loading. The center connected N-poles resonator consisting of crossed dipole arms with loading at the ends provides flexible and stable wide frequency response [9, 10]. Here the end loading on the crossed arms is done using curved metal strips to further reduce the size of the resonator. The geometry of the metasurface resonator is shown in Fig. 9.7; the resonator consists of two back-to-back Rogers RT5880 substrates (thickness of 0.254 mm), metal cross dipoles with curved metal end loading is printed on one of the faces of the substrate. The curved arms are folded inwards in order to achieve compact size to avoid phase errors in the array. The two dielectric slabs consisting of same patterns are separated by an air gap of $h = 0.8$ mm in-between to extend the transmission bandwidth up to 15 GHz as shown in the figure. The width of the cross-dipole arms is 0.67 mm and the length of curved end-loading is increased or decreased by the gap '$g$' to shift the transmission band to lower or higher frequency respectively. This way the transmission phase can be tuned to achieve any value required for the phase compensation. The designed lens constitutes number of transmission type metasurface unit cells with each unit cell contributing to a continuous transmission phase depending upon the gap '$g$'.

The unit cell is studied using EM simulations for the plane wave excitation with periodic boundary conditions on both sides. The transmission magnitude and phase

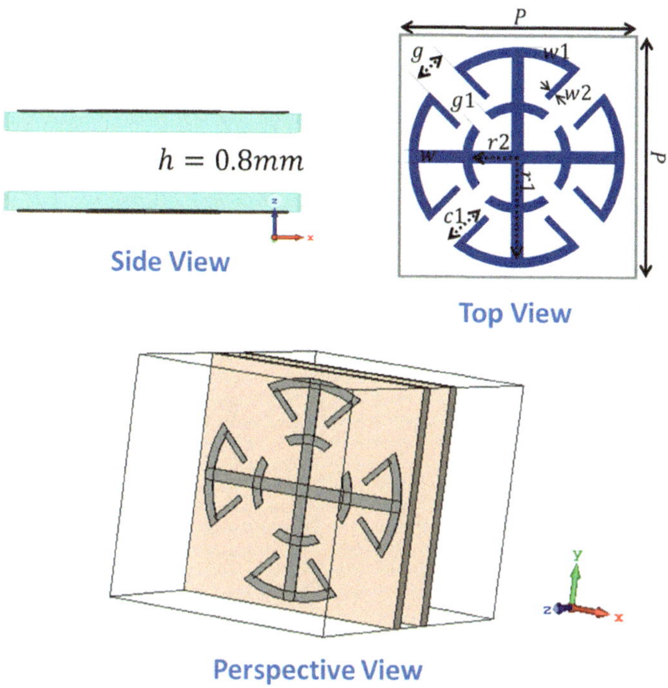


**Fig. 9.7** Geometry of the double layered MS unit cell consisting of two identical layers of substrate and metal rings ($p = 4$, $r_1 = 2 \times r_2 = 1.7$, $w = 0.2$, $w1 = 0.15$, $w2 = 0.1$, $c1 = 0.75$, $g1 = 0.4$, all dimensions are in millimetres). ([8], Reprinted with permission from IEEE)

are plotted for different values of $g$ in Fig. 9.8. By properly choosing the value of $g$ ranging from 0.1 to 1.40 mm, 0°–260° transmission phase is achieved (with $|S_{21}| > 0.4$), we can see a 15 GHz transmission bandwidth is achieved using the double layered MS unit cell. The unit cell has almost a linear phase response for variations in the gap $g$ as shown in Fig. 9.9.

### *9.4.1 Study of Lensing Property*

The planar ultra-thin lens is designed using the MS unit cells presented above, has an aperture size of $5.6\lambda_0$ (at 28 GHz) and has a thickness of $0.12\lambda_0$ resulting in focal-to-diameter ration, $f/D$ of 0.2. The configuration of the lens consists of 225 phase tuned unit cells (4 mm × 4 mm in size) and these cells are symmetric with respect to x- and y-axis. The phase delay required at each location given by (9.3) is compensated by properly choosing the gap ‘g’ of the unit cell. The lensing performance is studied using the full wave EM simulation as shown in Fig. 9.10a. The full wave simulations of the lens with the x-polarized plane wave incident on the lens from the bottom is

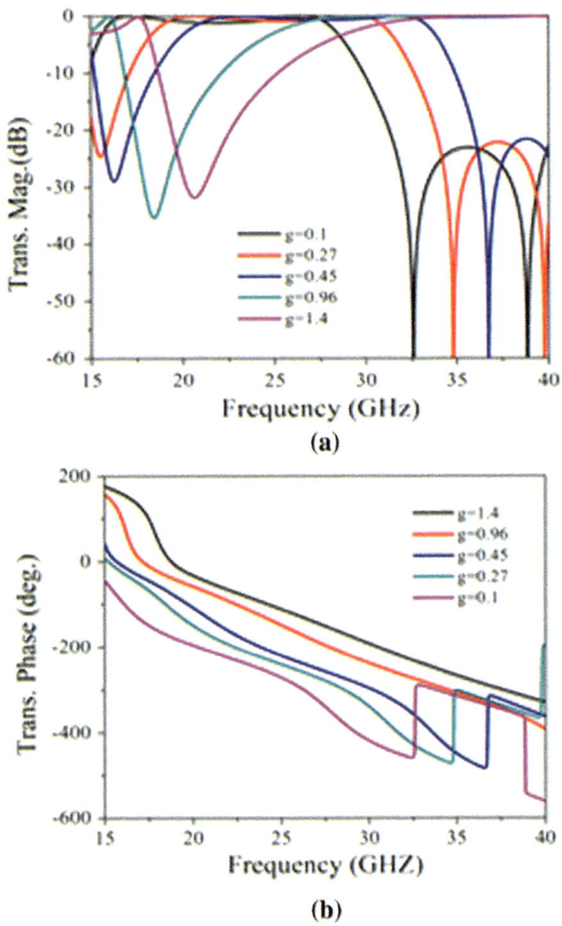


**Fig. 9.8** Performance of the MS unit cell for plane wave excitation **a** transmission magnitude versus gap $g$, **b** transmission phase versus gap $g$. ([8], Reprinted with permission from IEEE)

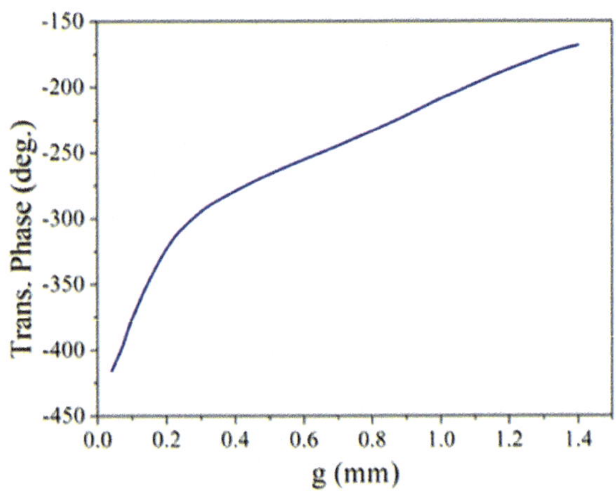


**Fig. 9.9** Transmission phase coverage versus gap $g$. ([8], Reprinted with permission from IEEE)

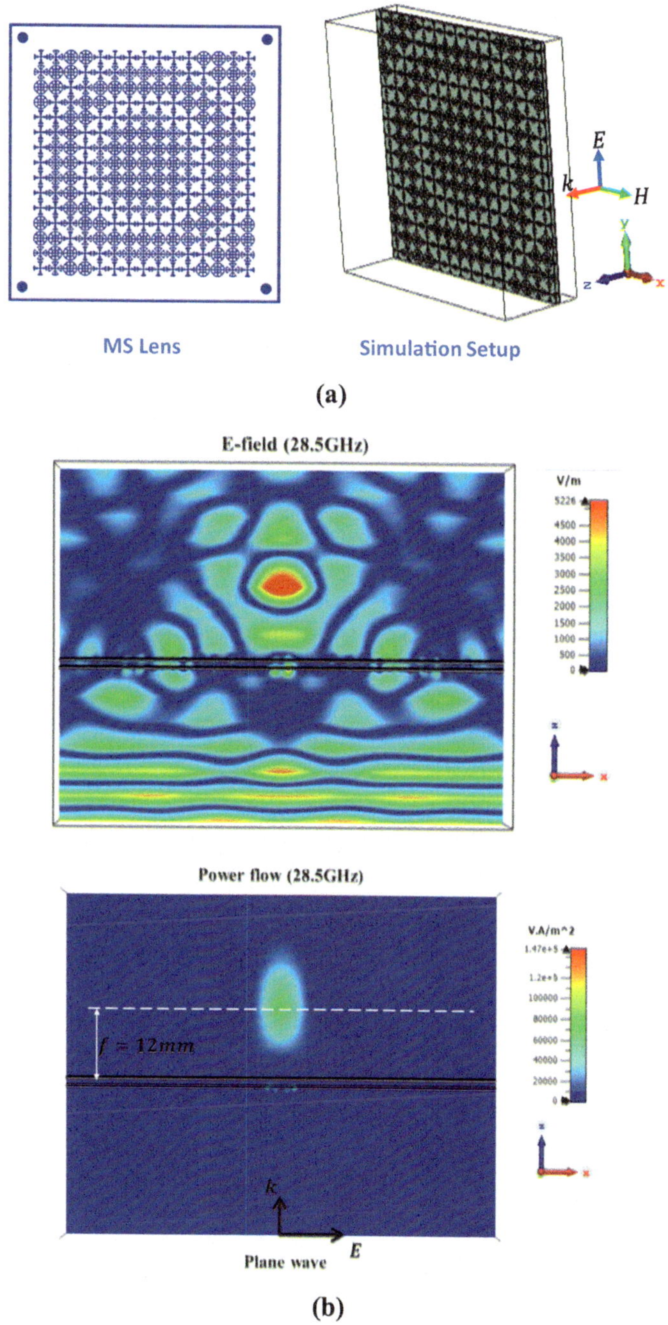


**Fig. 9.10** Thin planar lens EM study **a** lens top view and simulation setup and **b** electric field and power flow through the lens in xz-plane at 28.5 GHz. ([8], Reprinted with permission from IEEE)

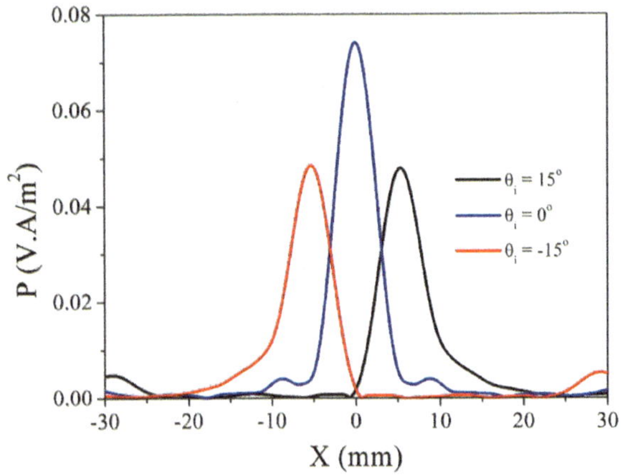


**Fig. 9.11** Power distribution at focal plane $z = 12$ mm along x-direction versus incident angles. ([8], Reprinted with permission from IEEE)

shown in Fig. 9.10b. We can observe that the converging response of the lens for a plane wave incident on other side. Most of the power is concentrated at the focal point of the lens which verifies the lensing property. To further study the lens property for oblique incidence, a plane wave is incident on the lens at $-15°$, 0° and 15°, the power distribution at the focal plane $z = 12$ mm on the other side is plotted along x-direction in Fig. 9.11. From the figure, we can observe that the power is focused at $x = -5.3$ mm, 0 and 5.3 mm for the three incident angles ($-15°$, 0° and 15°) of the plane wave, which confirms the lensing property of the metasurface.

## 9.5 Circularly Polarized Antenna Array

The three-antenna array with each antenna fed via proximity feed coupling printed on a Rogers RT5880 substrate with thickness, $h = 0.254$ mm placed below the top-substrate (Rogers RT 5880 and $h = 0.254$ mm) on which three patch antennas are printed, is shown in Fig. 9.12a. The tapered open ended microstrip line is positioned below the patch for the coupling of the fields to the patch [11]. In order to achieve wide impedance bandwidth a rectangular slot is made on the top patch which has chamfered corner to generate circular polarization. The antenna is optimized, and the various dimensions are given in the antenna configuration. The antennas are separated by 5.7 mm to avoid coupling of the energy between the ports for better MIMO performance. The antenna array is fabricated by printing feed lines and the patches on two different substrates and then bonding the two substrates together using the glue. The photograph of the antenna array is shown in Fig. 9.12b. The antenna array is tested for port performance and the simulated and measured S-parameters of the antenna plotted in Fig. 9.13. The aperture coupled antenna is designed to cover 27.5–31 GHz frequency band and the measured results confirm 28–31.9 GHzimpedance bandwidth. The isolation between the ports 1 and 2 (same for 2 and 3) is greater

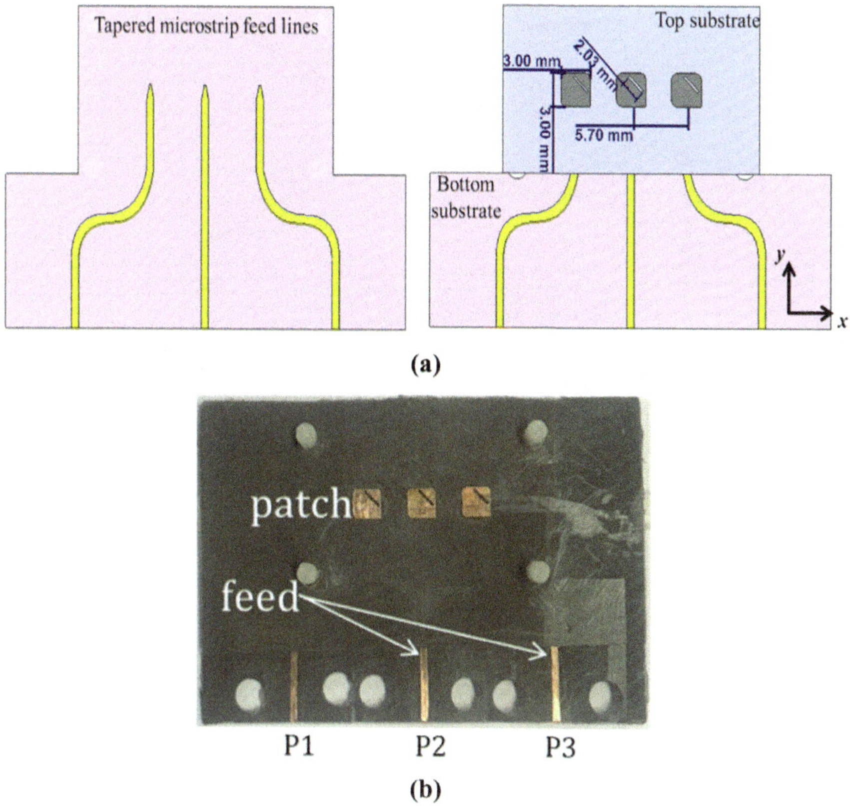


**Fig. 9.12** **a** Configuration of the proximity fed aperture coupled antenna array and **b** photograph of the fabricated three-antenna array. ([8], Reprinted with permission from IEEE)

than 22 dB in the band, which is quite good for MIMO performance. The right-hand circular polarization (RHCP) and left-hand circular polarization (RHCP) radiation patterns for the single aperture coupled patch antenna-2 is presented in Fig. 9.14a at 29.5 GHz. the radiation pattern for antennas 1 and 3 is the same and the antennas have circular radiation pattern. The gain and axial ratio versus the frequency is plotted in Fig. 9.14b, 3 dB axial ratio is achieved for 29.3–30.3 GHz frequency band. The antenna-2 has a gain of about 6.9dBi in this frequency band which quite low for mmWave applications.

## 9.6 Antenna Array with Planar Metasurface Lens

After the design of the antenna array, the ultra-thin planar lens designed in Sect. 9.4 is loaded on top of the antenna array at $f = 12$ mm and the radiation performance

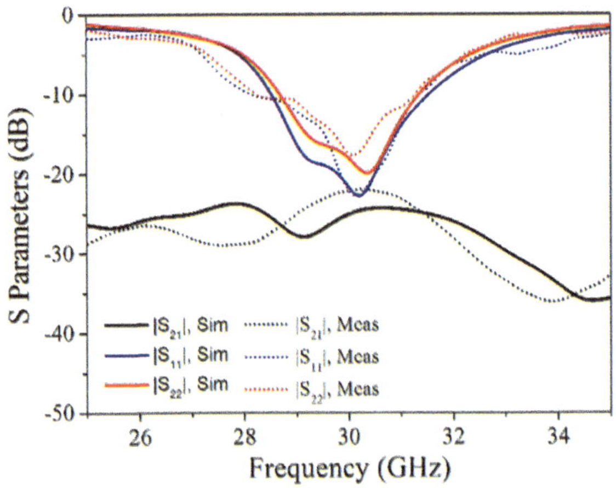


**Fig. 9.13** Measured and simulated S-parameters of the antenna. ([8], Reprinted with permission from IEEE)

is studied. The radiation performance of the antenna array with the lens is studied using the full wave EM simulations. The 3D radiation pattern of the antenna-2 with the planar lens is presented in Fig. 9.15 at 29 GHz, directivity of 20.25dBi with 11° and 12°beamwidths in two orthogonal planes is achieved with the MS lens. With the peak gain of 20.2 dB at 29 GHz, the peak aperture efficiency with the lens size of 60 mm × 60 mm is about 24.7%. To have a better understating of peak aperture efficiency of the MS lens, the aperture size is optimized for maximum aperture efficiency. With such a small focal length, all the MS unit cells in the larger aperture size are not properly excited thus resulting in maximum of 24.7% aperture efficiency only. Therefore, aperture size is reduced in simulations: the 3D radiation pattern and polar plot of the MS lens having an aperture size of 44 mm × 44 mm is shown in Fig. 9.16a. A peak gain of 18.9 dB is observed at 29 GHz for the centre element with the smaller aperture having size 1936 mm$^2$ (consisting of 11 × 11 unit cells) amounting to a peak aperture efficiency of 34%. The radiation patterns for port 1 and 2 are also shown in Fig. 9.16b, with the same aperture the radiation is pointed towards 15 degrees with a beam peak of 17 dB.

### *9.6.1 Implementation and Measured Results*

The MS lens is fabricated using the standard PCB fabrication process and consists of two layers separated by an air gap of 0.8 mm. The air gap between the two layers is maintained using the 3D printed stand and Rohacell foam in-between the two layers. The MS lens is fabricated using Rogers RT5880 laminates with thickness 0.254 mm, has an aperture area of 60 × 60 mm$^2$ and total thickness of the lens is 1.3 mm. The photograph of the lens mounted on the three-antenna array at focal

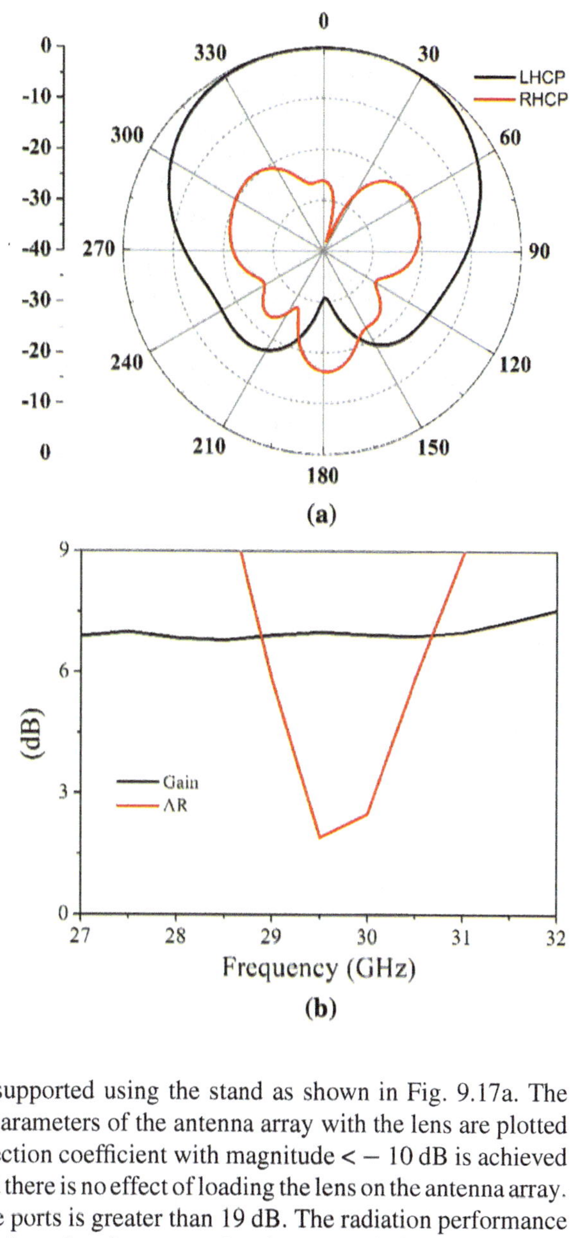


**Fig. 9.14** **a** Radiation pattern of the antenna-2 at 29.5 GHz and **b** gain and axial ratio of the antenna-2 versus frequency. ([8], Reprinted with permission from IEEE)

length of 12 mm which is supported using the stand as shown in Fig. 9.17a. The measured and simulated S-parameters of the antenna array with the lens are plotted in Fig. 9.17b. The input reflection coefficient with magnitude $< -10$ dB is achieved for 28–31.9 GHz verifies that there is no effect of loading the lens on the antenna array. The isolation between all the ports is greater than 19 dB. The radiation performance of the antenna array with the planar lens is measured and compared with the simulated results using a standard Ka-band horn antenna. The radiation patterns for the three antennas are presented in Fig. 9.18a–c. The patterns show that the beam points to 15°, 0° and −15° when antennas 1–3 are excited, which confirms the beam scanning properties of the planar lens. It is worth noting that the antennas are separated along the x-direction resulting in the beam forming at ±15° with respect to antenna-2

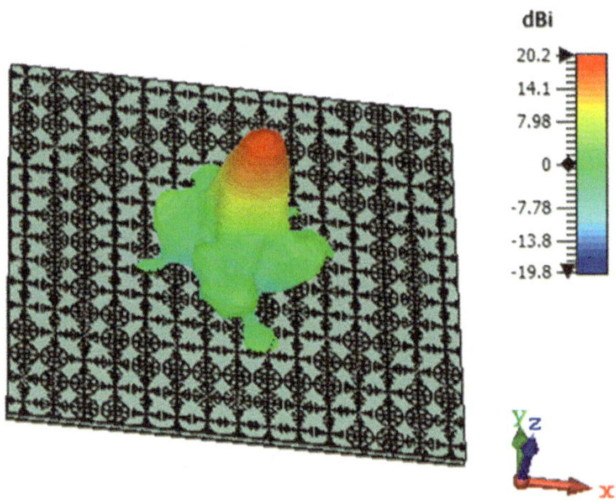


**Fig. 9.15** 3D radiation pattern of the antenna-2 with the thin planar lens at 29 GHz. ([8], Reprinted with permission from IEEE)

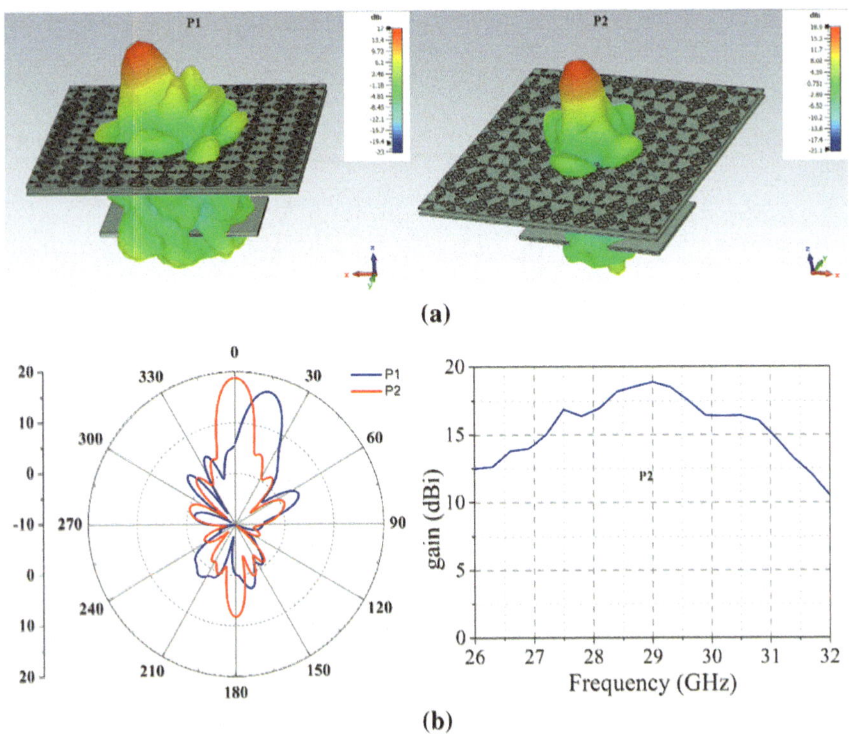


**Fig. 9.16** MS lens with smaller aperture **a** 3D pattern for P1 and P2, **b** radiation pattern and gain for P2

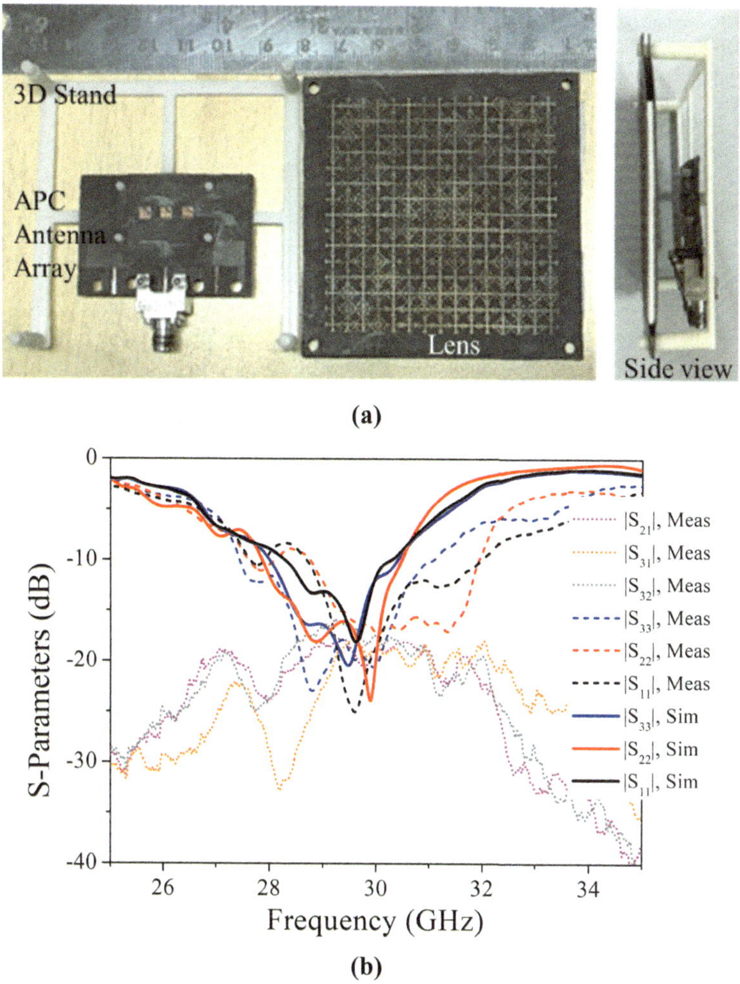


**Fig. 9.17** Antenna array with MS lens **a** photograph of the prototype antenna, **b** S-parameters of the antenna. ([8], Reprinted with permission from IEEE)

which we have also observed in planar lens study using the plane wave excitation. However, the offset excitation results in the side-lobe-levels of − 10 dB for the ports 1 and 3. Also, the front to back ratio (FBR) with the lens is 10 dB due to the smaller ground plane size with respect to the lens, but the FBR can be increased using a larger ground plane. To verify the lens polarization insensitivity, the co- and cross-polarized radiation patterns for the antenna-2 are measured and the same are plotted in Fig. 9.19 at 30 GHz. The co- and cross-polarization radiation measured has the same magnitude which confirms that the planar lens can work with any polarization.

The simulated and measured gain is plotted in Fig. 9.20 as a function of the frequency, a peak gain of 20.2 dBi at 29 GHz is achieved using the lens over the

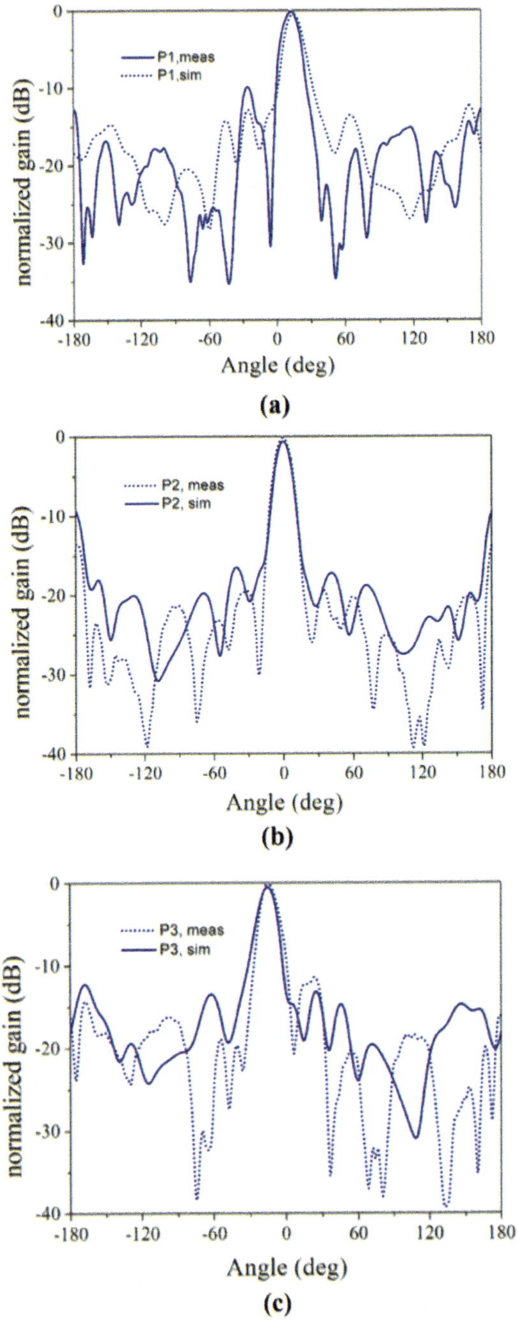


**Fig. 9.18** Measured and simulated input reflection coefficient of the patch antenna with the MTM lens loading, **a** Port P1, **b** Port P2 and **c** Port P3. ([8], Reprinted with permission from IEEE)

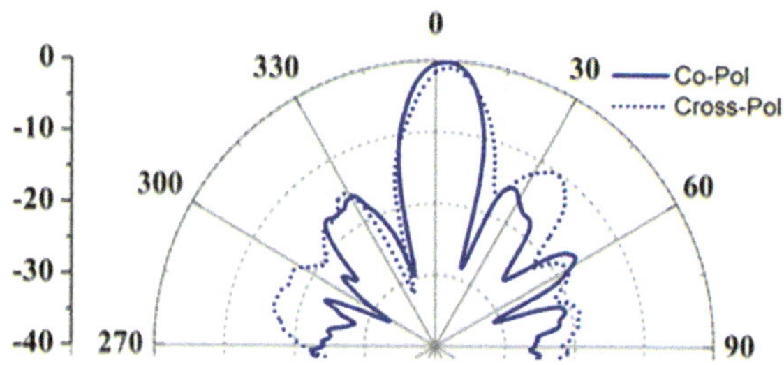


**Fig. 9.19** Measured co- and cross-pol radiation pattern for antenna-2 with the MS lens at 30 GHz. ([8], Reprinted with permission from IEEE)

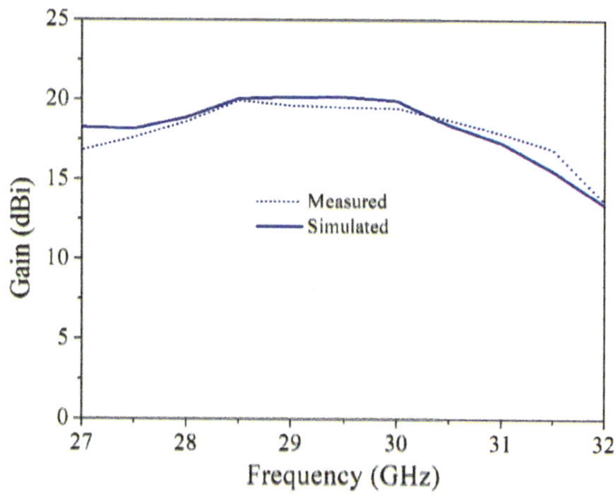


**Fig. 9.20** Measured and simulated realized gain of the antenna-2 with planar lens. ([8], Reprinted with permission from IEEE)

antenna array. With the peak gain of 20.2 dBi for the aperture size of 36 $cm^2$, the calculated aperture efficiency is about 24.7%. Therefore, with the planar MS lens, one can achieve a beam scanning of $-15°$ to $+15°$ at three ports.

The comparison in Table 9.1 shows the difference between GRIN lens and the PGMS lens to highlight the features of the planar PGMS lens antenna at mmWave frequency band. The antenna array with PGMS lens has a very compact profile and can be easily extended to large antenna arrays to achieve wide beam scanning. The lens can be translated to any frequency band with sufficient bandwidth and does not disturb the antenna impedance due to matched unit cells. A peak aperture efficiency of 34% is achieved with 3 dB gain BW (GBW) above 10% for $f/D = 0.27$. This aperture efficiency is due to the smaller unit cell design with high transmission properties. Also, it is seen that if the designer wants to decrease the unit cell size further, he must increase the thickness of the unit cells (multilayer) for the same phase coverage. However, the multilayer design at mmWaves is practically not feasible due to losses involved. Most of the references utilize multilayer (more than two substrates) phase unit cells in above table for phase coverage.

**Table 9.1** Comparison of different lens antenna arrays

| Refs. | f(GHz) | Lens aperture/thickness | Focal length | AE (%) | 3 dB GBW | Beam scanning | Remarks |
|---|---|---|---|---|---|---|---|
| [7] | 9.5 (6.7–12) | 20.79 $\lambda^2$/0.38 $\lambda$ | (3.2 $\lambda$) 100.8 mm | 50 | 61% (7–13.2 GHz) | N/A | GRIN MS lens fed via single tapered slot antenna |
| [12] | 10 | 18.77$\lambda^2$/0.15$\lambda$ | (1.16 $\lambda$) 35 mm | 29.9 | N/A | N/A | PGMS lens fed via single MPA |
| 3-port lens antenna array [8] | 28 (27.5–32) | 31.3$\lambda^2$/0.12$\lambda$ | 1.12 $\lambda$ 12 mm | 24.7 | 12%(27.5–31 GHz) | ±15° | PGMS lens fed via three AP antenna array (f/D = 0.2) |
| | | 16.8$\lambda^2$/0.12$\lambda$ f/D = 0.27 | 1.12 $\lambda$ 12 mm | 34 | 10.3% (27.5–30.5 GHz) | ±15° | PGMS lens fed via three AP antenna array (f/D = 0.27) |

## 9.7 Conclusion

This chapter presents lens antenna arrays for mmWave MIMO applications. The implementation of mmWave MIMO technique using lens antenna arrays is discussed. The lens when implemented with antenna arrays results in significant reduction in hardware cost and system complexity for mmWave MIMO applications. Therefore, the design of thin planar EM lenses is quite crucial. For the design of metasurface lens, two techniques such as GRIN and PGMS are discussed in detail. Both these techniques are compared in this chapter from design and performance perspectives. Lastly, a three-port antenna array with the PGMS lens is presented for 28 GHz MIMO applications.

## References


1. Alkhateeb A, Mo J, Gonzalez-Prelcic N, Heath RW (2014) MIMO Precoding and combining solutions for millimeter-wave systems. IEEE Commun Mag 52(12):122–131. https://doi.org/10.1109/MCOM.2014.6979963
2. Roh W et al (2014) Millimeter-wave beamforming as an enabling technology for 5G cellular communications: theoretical feasibility and prototype results. IEEE Commun Mag 52(2):106–113. https://doi.org/10.1109/MCOM.2014.6736750
3. Sun S, Rappaport TS, Heath RW, Nix A, Rangan S (2014) Mimo for millimeter-wave wireless communications: beamforming, spatial multiplexing, or both? IEEE Commun Mag 52(12):110–121. https://doi.org/10.1109/MCOM.2014.6979962
4. Sayeed A, Behdad N (2010) Continuous aperture phased MIMO: basic theory and applications. In: 2010 48th annual allerton conference on communication, control, and computing (Allerton), pp 1196–1203. https://doi.org/10.1109/ALLERTON.2010.5707050
5. Brady J, Behdad N, Sayeed AM (2013) Beamspace MIMO for millimeter-wave communications: system architecture, modeling, analysis, and measurements. IEEE Trans Antennas Propag 61(7):3814–3827. https://doi.org/10.1109/TAP.2013.2254442
6. Zeng Y, Zhang R (2016) Millimeter wave MIMO with lens antenna array: a new path division multiplexing paradigm. IEEE Trans Commun 64(4):1557–1571. https://doi.org/10.1109/TCOMM.2016.2533490
7. Erfani E, Niroo-Jazi M, Tatu S (2016) A high-gain broadband gradient refractive index metasurface lens antenna. IEEE Trans Antennas Propag 64(5):1968–1973. https://doi.org/10.1109/TAP.2016.2526052
8. Wani Z, Abegaonkar MP, Koul SK (2021) Thin planar metasurface lens for millimeter wave MIMO applications. IEEE Trans Antennas Prop. https://doi.org/10.1109/TAP.2021.3098571
9. Munk BA (2005) Frequency selective surfaces: theory and design. Wiley
10. Volakis JL, Sertel K (2011) Narrowband and wideband metamaterial antennas based on degenerate band edge and magnetic photonic crystals. Proc IEEE 99(10):1732–1745. https://doi.org/10.1109/JPROC.2011.2115230
11. Pozar D, Jackson R (1987) An aperture coupled microstrip antenna with a proximity feed on a perpendicular substrate. IEEE Trans Antennas Propag 35(6):728–731. https://doi.org/10.1109/TAP.1987.1144167
12. Li H, Wang G, Xu H, Cai T, Liang J (2015) X-band phase-gradient metasurface for high-gain lens antenna application. IEEE Trans Antennas Propag 63(11):5144–5149. https://doi.org/10.1109/TAP.2015.2475628

# Index





S. K. Koul and Z. Wani, *Novel Millimetre Wave Antennas for MIMO and 5G Applications*, Lecture Notes in Electrical Engineering 819,
https://doi.org/10.1007/978-981-16-7278-1

The manufacturer's authorised representative in the EU is Springer Nature Customer Service Centre GmbH, Europaplatz 3, 69115 Heidelberg, Germany. If you have any concerns regarding our products, please contact ProductSafety@springernature.com

Printed and bound by CPI Group (UK) Ltd, Croydon, CR0 4YY
30/09/2026
02225265-0013